O9-BTJ-738

Also by James Burke

Connections

The Day the Universe Changed

The Axemaker's Gift
(with Robert Ornstein)

The Pinball Effect

The Knowledge Web

From Electronic Agents to Stonehenge and Back—and Other Journeys Through Knowledge

JAMES BURKE

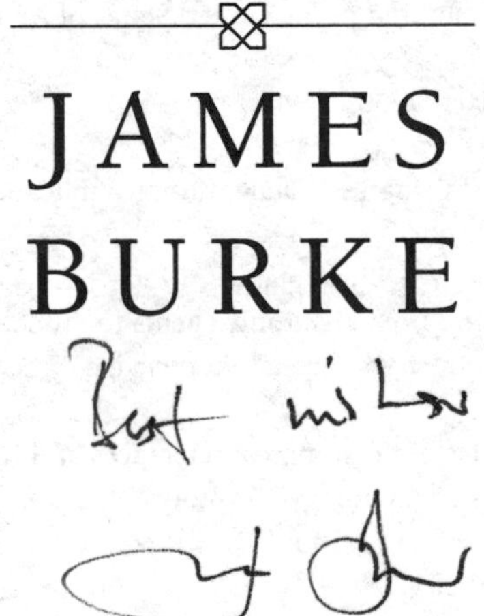

Simon & Schuster

SIMON & SCHUSTER
Rockefeller Center
1230 Avenue of the Americas
New York, NY 10020

Designed by Pagesetters

Manufactured in the United States of America

10 9 8 7 6 5 4 3 2 1

Library of Congress Cataloging-in-Publication Data

Burke, James, 1936–
The knowledge web : from electronic agents to Stonehenge and back—
and other journeys through knowledge / James Burke.
p. cm.
Includes bibliographical references and index.
1. Technology—History. I. Title.
T15.B763 1999
609—dc21 99-24539
CIP

ISBN 0-684-85934-3

To Madeline

Acknowledgements

I should like to thank Carolyn Doree and Jay Hornsby for their extremely valuable assistance in research.

Contents

Introduction

Change comes so fast these days that the reaction of the average person recalls the depressive who takes some time off work and heads for the beach. A couple of days later his psychiatrist gets a postcard from him. The message on the card reads: "Having a wonderful time. Why?"

Innovation is so often surprising and unexpected because the process by which new ideas emerge is serendipitous and interactive. Even those directly involved may be unaware of the outcome of their work. How, for instance, could a nineteenth-century perfume-spray manufacturer and the chemist who discovered how to crack gasoline from oil have foreseen that their products would come together to create the carburetor? In the 1880s, without the accidental spillage of some of the recently invented artificial colorant onto a petri-dish culture that revealed to a German researcher named Ehrlich that the dye preferentially killed certain bacilli, would Ehrlich have become the first chemotherapist? If the Romantic movement's concept of "nature-philosophy" had not suggested that nature evolves through the reconciliation of opposing forces, would Oersted have sought to "reconcile" electricity and magnetism and discovered the electromagnetic force that made possible modern telecommunications?

Small wonder, then, that the man and woman in the street are left behind in all this, if the researchers themselves don't get the point. But given the conditions under which science and technology work, how else could it be? At last count there were more than twenty thousand different disciplines, each of them staffed by researchers straining to replace what they produced yesterday.

These noodling world-changers are spurred on by at least two powerful motivators. The first is that you are more likely to achieve recognition if you make your particular research niche so specialist that there's only room in it for you. So the aim of most scientists is to know more and more about less and less, and to describe what it is they know in terms of such precision as to be virtually incomprehensible to their colleagues, let alone the general public.

The second motivator is the CEO. Corporations survive in a changing world only by encouraging their specialists to generate change before somebody else does. Winning in the marketplace means catching the competition by surprise. Not surprisingly, this process also surprises the consumer, and nowhere so frequently today as in the world of electronics, where by the time the user gets around to reading the manual, the gizmo to which it refers is obsolete.

We live in this permanently off-balance manner because of the way knowledge has been generated and disseminated for the last 120,000 years. In early Neolithic times the requirement to teach the highly precise, sequential skills of stone-tool manufacture demanded a similarly precise, sequential use of sounds and is thought to have given rise to language. The sequential nature of language facilitated description of the world in similarly precise terms, and in due course a process originally developed for chipping pieces off stone became a tool for chipping pieces off the universe. This reduction of reality to its constituent parts is at the root of the view of knowledge known as "reductionism," from which science sprang in the seventeenth-century West. Simply put, scientific knowledge comes as the result of taking things apart to see how they work.

Over millennia, this way of doing things has tended to subdivide knowledge into smaller and more specialist segments. For example, in the past hundred years or so, the ancient discipline of botany has fragmented and diversified to become biology, organic chemistry, histology, embryology, evolutionary biology, physiology, cytology, pathology, bacteriology, urology, ecology, population genetics and zoology.

There is no reason to suppose that this process of proliferation and fragmentation will lessen or cease. It is at the heart of what, since Darwin's time, has been called "progress." If we live today in the best of all possible materialist worlds, it is because of the tremendous strides made by specialist research that have given us everything from more absorbent diapers to linear accelerators. We in the technologically advanced nations are healthier, wealthier, more mobile, better-informed individuals than ever before in history, thanks to myriad specialists and the products of their pencil-chewing efforts.

However, the corollary to a small minority knowing more and more about less and less is a large majority knowing less and less about more and more. In the past this has been a relatively unimportant matter principally because for most of history the illiterate majority (hard-pressed enough just to survive) has been unaware that the problem existed at all. Technology was in such limited supply that there was only enough to share it among a few elite decision-makers.

It is true that over time, as the technology diversified, knowledge slowly diffused outward into the community via information media such as the alphabet, paper, the printing press and telecommunications. But at the same time these systems also served to increase the overall amount of specialist knowledge. What reached the general public was usually either out-of-date or no longer vital to the interests of the elite. And as specialist knowledge expanded, so did the gulf between those who had information and those who did not.

Each time there was a major advance in the ability to generate, store or disseminate knowledge, it was followed by an "information surge" and with it a sudden acceleration in the level of innovation that dramatically enhanced the power of the elites. But sooner or later the same technology reached enough people to undermine the status quo. The arrival of paper in thirteenth-century Europe strengthened the hand of church and throne, but at the same time created a merchant class that would ultimately question their authority. The printing press gave Rome the means to enforce obedience and conformity, then Luther used it to wage a propaganda war that ended with the emergence of Protestantism. In the late nineteenth century, when military technology made possible conflicts in which hundreds of thousands died, and manufacturing technology generated untenable working and living conditions for millions of factory workers, radicals and reformers were aided in their efforts by new printing techniques cheap enough to spread their message of protest in newspapers and pamphlets.

By the mid twentieth century scientific and technological knowledge far outstripped the ability of most people, even the averagely well-informed, to comprehend it. The stimulus of the Cold War brought advances in computer technology that seemed likely to place unprecedented power in the hands of economic and political power blocs. There was talk of "Big Brother" government, rule by multinational corporations, the central databases that would hold personal files on every individual, and the creeping homogenization of the human race into one giant "global village." Unchecked state and corporate industrialization finally began to generate the first visible signs of global warming, runaway pollution decimated the animal population and the tropical forests went down before fire and axe at an alarming rate.

However, at the same time, the falling cost of computer and telecommunications technology also began to make it possible for these developments to be discussed in an unprecedentedly large public forum. And the more we learned about the world through television and radio, the more it became clear that urgent measures were needed to preserve its fragile ecosystems and its even more fragile cultural diversity. At the end of the twentieth century the emergence of the ubiquitous Internet and affordable wireless technology offered the opportunity for millions of individuals to think of becoming involved.

However, the culture of scarcity with which we have lived for millennia has not prepared us well for the responsibilities technology will force on us in the next few decades. Reductionism, representative democracy and the division of labor have tended to leave such matters in the hands of specialists who are, increasingly, no more aware of the ramifications of their work than anybody else.

The result is that national and international institutions are coming under unprecedented stress as they try to apply their obsolete mechanisms to twenty-first-century problems. In Britain recently a case was brought against an individual which rested on the fifteenth-century meaning of the word "obscene." Medical etiquette has changed little since 1800. In some places science and religion are in conflict over the definition of life.

Western institutions function as if the world had not changed since they were established to deal with the specific problems of the time. Fifteenth-century nation-states, emerging into a world without telecommunications, developed representative democracy; seventeenth-century explorers in need of financial backing invented the stock market; in the eleventh century the influx of Arab knowledge triggered the invention of universities to process the new data for student priests.

In the coming decades it is likely that many social institutions will attempt to adapt by becoming virtual, bringing their services directly to the individual much in the way that banks have already begun to. But their new accessibility will in turn likely subject them to proliferating and diversifying demands that will change how they work and make them redefine their purpose. In education, the old reductionist reliance on specialism and testing by repetition will have to give way to a much more flexible definition of ability. As machines increasingly take over the tasks that once occupied a human lifetime, specialist skills may take on a merely antiquarian value. New ways will have to be found to assess intelligence in a world in which memory and experience seem no longer of value (again, this is nothing new: the alphabet and later the printing press both presented the same perceived threat).

When a corporate workforce becomes scattered across the country, or the globe, in thousands of individual homes or groups, and deals direct with millions of customers, the value of communication skills is likely to outweigh that of most others. Such ability may be possessed by people who would previously have been thought unqualified to work for the corporation, because in the old world they would have been too young, or too old, or too distant, for example. A virtual education system will have to deal with problems such as a multicultural global student body bringing very diverse experience, attitudes and aims to the class. In terms of international law, recent cases involving copyright or pornography reveal how complex such legal problems are likely to become.

This book does not attempt directly to address any of these problems. Rather, it suggests an approach to knowledge perhaps more attuned to the needs of the twenty-first century as described above. Some readers will no doubt see this approach as more evidence of the "dumbing-down" of recent years. But the same was said about the first printing press, newspapers, calculators and the removal of mandatory Latin from the curriculum.

In its fully developed form, the "webbed" knowledge system introduced here would be inclusive, not exclusive. Modern interactive networked communications systems married to astronomically large data storage capability ought to ensure that at times of change nothing need be lost. No subject or skill will be too arcane for its practitioners to pursue when the marketplace for their skills is planetwide.

Also, no external memory device from alphabet to laptop seems to have degraded human mental abilities by its introduction. Rather these abilities have been augmented each time by the new tools. Some skills, such as rote memory, become less widely used, but there seems to be no evidence that the capability for them disappears. In many cases machines also take over routine work, freeing individuals to use their skills at higher levels.

The latest interactive, semi-intelligent technologies seem likely to make this possible on an unprecedented scale. They also bring to an end a period of history in which the human brain was constrained by limited technology to operate in a less-than-optimal way, since the brain appears not to be designed to work best in the linear, discrete way promoted by reductionism. The average healthy brain has more than a hundred billion neurons, each communicating with others via thousands of dendrites. The number of potential ways for signals to go in the system is said to be greater than the number of atoms in the universe. In matters as fundamental as recognition it seems that the brain uses some of its massive interconnectedness to call on many different processes at once to deal with events in the outside world, so as quickly to identify a potentially dangerous pattern of inputs.

It is this pattern-recognition capability that might prove to be the most useful attribute of a webbed knowledge system driven by the semi-intelligent interactive systems now being developed. As this book hopes to show, learning to identify the pattern of connections between ideas, people and events is the first step toward understanding the context and relevance of information. So the social implications of webbed knowledge systems are exciting, since they will make it easier for the average citizen to become informed of the relative value of innovation. After all, it is not necessary to understand the mathematics of radioactive decay to make a decision about where to site a nuclear power plant. As I hope you will see, this approach to knowledge may be one way to enfranchise those millions who lack what used to be called formal education and to move us toward more participatory forms of government.

I would not pretend that what follows is more than a first exercise, a number of linked storylines intended to introduce the reader to the kind of information infrastructures we may begin to use in the next few decades. But I hope they will introduce the reader to a new, more relevant way of looking at the world, because in one way or another, we're all connected.

JAMES BURKE
London 1999

How to Use This Book

The Knowledge Web takes ten different journeys across the great web of change. There are many different ways to read this book, just as there are many different ways to travel on a web. The simplest way is to read from start to finish, in the manner unchanged since the appearance of alphabetic writing thirty-five hundred years ago. Or you can read the book the way your teacher once told you not to. You can do this at many points throughout the book, when the timeline of a particular journey reaches a "gateway" on the web, where it crosses with the timeline of another, different journey. At such a gateway, you'll see the coordinates for the location of that other place.

Using the coordinates you may, if you choose, jump backward or forward (through literary subspace) to the other gateway, pick up the new timeline and continue your journey on the web, until you reach yet another gateway, when you may, if you choose, jump once again. The coordinates that identify a gateway appear in the text like this:

> This was the ducal seat of the Hamilton family and leased by Dr. John **17** 36 *68*
> Roebuck,[17] a successful entrepreneur and ex-pupil of Black's.

In the text, "Roebuck,[17]" is the site of the seventeenth gateway so far. In the margin, "36 *68*" is the gateway you'll jump to (the thirty-sixth gateway, located on page 68).

Sometimes, there'll be multiple gateways you can jump to, at busy moments in history, where several pathways of change meet. Good luck!

Since there are 142 gateways here that cross, then in one sense, that means this book could conceivably be read at least 142 different ways. Though I don't suggest you try it, doing so would give you a rather visceral feeling for the way change happens.

And it happens that way to all of us, all the time. It's happening to you now, though you may not know it yet.

CHAPTER 1

Feedback

This book takes a journey across the vast, interconnected web of knowledge to offer a glimpse of what a learning experience might be like in the twenty-first century once we have solved the problem of information overload.

In the past when technology generated information overload the contemporary reaction was much the same as it is today. On the first appearance of paper in the medieval West, the English bishop Samson of St. Alban's complained that because paper would be cheaper than animal-skin parchment people would use paper to write too many words of too little value, and since paper was not as durable as parchment, paper-based knowledge would in the long run decay and be lost. When the printing press was developed in the fifteenth century it was said that printed books would make reading and writing "the infatuation of people who have no business reading and writing." Samuel Morse's development of the telegraph promised to link places as far apart as Maine and Texas, triggering the reaction: "What have Maine and Texas to say to each other?" The twentieth-century proliferation of television channels has led to concerns about "dumbing-down."

The past perception that new information technologies would have a destabilizing social effect led to the imposition of controls on their use. Only a few ancient Egyptian administrators were permitted to learn the skills of penmanship. Medieval European paper manufacture was strictly licensed. The output of sixteenth-century printing presses was subject to official censorship by both church and state. The new seventeenth-century libraries were not open to the public. Nineteenth-century European telegraphs and telephones came under the control of government ministries.

The problem of past information overload has generally been of concern only to a small number of literate administrators and their semiliterate masters. In contrast, twenty-first-century petabyte laptops and virtually free access to the Internet may bring destabilizing effects of information overload that will operate on a scale and at a rate well beyond anything that has happened before. In the next few decades hundreds of millions of new users will have no experience in searching the immense amount of available data and very little training in what to do with it. Information abundance will stress society in ways for which it has not been prepared and damage centralized social systems designed to function in a nineteenth-century world.

Part of the answer to the problem may be an information-filtering system customized to suit the individual. The most promising of the systems now being developed will guide users through the complex and exciting world of information without their getting lost. This book provides an opportunity for the reader to take a practice run on such a journey. The journey (the book) begins and ends with the in-
1 142 *262* vention of the guidance system itself—the semi-intelligent agent.[1]

There are several types of agent in existence acting like personal secretaries in a variety of simple ways: filtering genuine e-mail from spam, running a diary, paying bills and selecting entertainment. In the near future agents will organize and conduct almost every aspect of the individual's life. Above all they will journey across the knowledge webs to retrieve information, then process and present it in ways customized to suit the user. In time they will act on behalf of their user because they will have learned his or her preferences by learning from the user's daily requirements.

In the search to develop semi-intelligent agents, one of the most promising systems (and the one which starts this journey) may be the neural network. Such a network consists of a number of cells each reacting to signals from a number of other cells that in turn fire their signals in reaction to input from yet other cells. If input signals cause one cell to fire more frequently than others, its input to the next cell in the series will be given greater weighting. Since cells are programmed to react preferentially to input from cells that fire frequently rather than from those that fire rarely, the system "learns" from experience. This is thought to be similar to the way learning operates in the human brain, where the repetition of a signal generated in response to a specific experience can cause enlargement in the brain cell's synapses.

The synapse is the part of the cell that releases transmitter chemicals that cross the gap to the next cell. If sufficient chemicals arrive on the other side, they generate an impulse. If enough of these signals are generated in the target cell, they cause its synapses to release chemicals in turn, and "pass the message on." A cell with larger synapses, releasing larger amounts of chemical, is therefore more likely to cause another cell to fire. Networks of such frequently firing cells may constitute the building blocks of memory.

This theory of neuronal interaction was first proposed in 1943 by two American researchers, Walter Pitts and Warren McCulloch, who also suggested that such a feedback process might result in purposive behavior when linking the senses with the brain and muscles if the result of the interaction were to cause the muscles to act to reduce the difference between a condition in the real world as perceived by the senses and the condition as desired by the brain.

Pitts and McCulloch belonged to a small group of researchers calling itself the "Teleological Society," another of whose members was the man who invented the name for this neural feedback process. He was Norbert Wiener, and he was the first to see the way in which feedback might work in a machine, during his research on antiaircraft artillery systems during World War II. Wiener was a rotund, irascible, cigar-chomping MIT professor of math who prowled what he described as the "frontier areas" between the scientific disciplines. Between biology and engineering Wiener developed a new discipline
to deal with feedback processes. He called the new discipline "cyber- **2** 141 *262*
netics."[2] Wiener recognized that feedback devices are information-processing systems receiving information and acting upon it. When

applied to the brain this new information-oriented view was a fundamental shift away from the entirely biological paradigm that had ruled neurophysiology since Freud, and it was to affect all artificial-intelligence work from then on.

Wiener first applied his feedback theory early in World War II, when he and a young engineer named Julian Bigelow were asked to improve the artillery hit rate. At the beginning of the war the problem facing antiaircraft gunners was that as the speed of targets increased (thanks to advances in engine and airframe technology) it became necessary to be able to fire a shell some distance ahead of a fast-moving target in order to hit it. Automating this process involved a large number of variables: wind, temperature, humidity, gunpowder charge, length of gun barrel, speed and height of target, and many others. Wiener used continuous input from radar tracking systems to establish the recent path of the target and use that path to predict what the target's likely position would be in the immediate future. This information would then be fed to the gun-moving mechanisms so that aiming-off was continually updated.

The system had its most outstanding successes in 1944, when British and American gunners shot down German flying bombs with fewer than one hundred rounds per hit. This was an extraordinary advance over previous performance, estimated at one hit per twenty-five hundred rounds. In 1944, during the last four weeks of German V-1 missile attacks on England, the success rate improved dramatically. In the first week, 24 percent of targets were destroyed; in the second, 46 percent; in the third, 67 percent; and in the fourth, 79 percent. The last day on which a large number of V-1s were launched at Britain, 104 of the missiles were detected by early-warning radar, but only four reached London. Antiaircraft artillery destroyed sixty-eight of them.

Early in his work on the artillery project Wiener had frequent discussions with a young physiologist named Arturo Rosenbleuth, who was interested in human feedback mechanisms that act to ensure precision in bodily movement. For the previous fifteen years Rosenbleuth had worked closely with Walter Cannon, professor of physiology at Harvard. Earlier in the century Cannon had invented the barium meal, which was opaque to X-rays. When ingested by a goose the barium revealed the peristaltic waves that occurred in the bird's stomach when it was hungry. Cannon observed that hunger seemed to precipitate the onset of these waves. He then observed that when a hungry animal was frightened the waves stopped.

This led to Cannon's ground-breaking studies of the physical effects of emotion. He discovered that when an animal was disturbed its sympathetic nervous system secreted into the bloodstream a chemical that Cannon named "sympathin." This chemical counteracted the effects of the disturbance and returned the animal's body systems to a state of balance. Cannon named the balancing process "homeostasis." In 1915 Cannon discovered that the principal body changes affected by the sympathetic system were those involved in fight, sexual activity or flight. In such situations sugar flowed from the liver to provide emergency energy and blood shifted from the abdomen to the heart, lungs and limbs. If the body were wounded, blood clotting occurred more rapidly than usual. In 1932 Cannon published a full-scale account of his research titled *The Wisdom of the Body*.

What had initially triggered Cannon's interest in homeostatic mechanisms was the work of the man to whom Cannon dedicated the French edition of his book. He was an unprepossessing but eminent French physiologist named Claude Bernard, who had started his working life as a pharmacist's assistant in Beaujolais, where his father owned a small vineyard. After being forced to give up his early schooling for lack of funds, Bernard took up writing plays. He produced first a comedy and then a five-act play, which he took to Paris in 1834 with the intention of making a career in the theater. Fortunately for the future health of humankind Bernard was introduced to an eminent theatrical critic, Saint-Marc Girardin, who read the play and advised Bernard to take up medicine.

At first Bernard planned to be a surgeon, but becoming dissatisfied with the general lack of physiological data he began to gather his own data by experimenting on animals. By 1839 his dexterity in dissection had brought him to the attention of the great physiologist François Magendie, who appointed him as assistant. One winter morning in 1846 some rabbits were brought to Magendie's lab for dissection and Bernard noticed that their urine was clear and acidic. As every nineteenth-century French winemaker knew, the urine of rabbits is usually turbid and alkaline. Bernard realized that the rabbits had not been fed and theorized that since the urine of carnivores is clear, the hungry, herbivorous rabbits must have been living on their fat. When he fed grass to the rabbits their urine returned to its normal alkaline turbidity. He double-checked with an experiment on himself. After twenty-four hours subsisting only on potatoes, cauliflower, carrots, green peas, salad, and fruit, Bernard's own urine went turbid and alkaline. Bernard then starved the rabbits, fed them boiled beef and dissected them to find out what had happened. He saw a milklike substance (he took it to be emulsified fat) that had formed at the point where the rabbit's pancreatic juice was pouring into the stomach. There was clearly some link between the juice and the emulsification of the fats.

Two years later he discovered the glycogenic function of the liver, which injects glucose into the blood. It was this discovery that led to Bernard's greatest contribution to the sum of human knowledge, because he saw that the function of the liver and the pancreas (and perhaps other systems, too) was to maintain the body's equilibrium. He summed up his research: "All the vital mechanisms, however varied they may be, have only one object, that of preserving constant the conditions of life in the inner environment." Follow-up research on the pancreas led an English researcher, William Bayliss, to coin the phrase that Cannon would use as his book title: "the wisdom of the body."

Not everybody was happy with Bernard's work, especially when he designed an oven in which to cook animals alive. An American doctor, Francis Donaldson, who attended Bernard's lectures in 1851, wrote: "It was curious to see walking about the amphitheater of the College of France dogs and rabbits, unconscious contributors to science, with five or six orifices in their bodies from which at a moment's warning, there could be produced any secretion of the body, including that of the several salivary glands, the stomach, the liver, and the pancreas."

Bernard was well aware of public opposition to vivisection but defended it: "The science of life is like a superb salon resplendent with light which one can enter only through a long and ghastly kitchen." Alas, Bernard's wife was unable to take the heat. After leaving him in 1869 she went in search of the antivivisection activists to whom she had been sending regular contributions.

She did not have far to go. In Paris a fanatical young vegetarian Englishwoman named Anna Kingsford, the owner of the *Lady's Own Paper,* had come to France to study medicine. Kingsford became well-known at the medical school for refusing to let her professors vivisect during the lessons she attended and for demonstrating against the practice. Kingsford's lecture halls were close to Bernard's labs, and she became so obsessed by his work that she set about directing all her energies toward killing him with thought waves. Bernard died only few weeks after she had begun to concentrate her mental energies on him, convincing her that she had been the instrument of divine will. Kingsford also claimed to have been responsible for the death of another vivisector, Paul Bert. However her efforts to
do the same to Louis Pasteur[3] failed. 3 54 *114*

Legislation to protect animals from ill treatment took a long time 3 125 *235*
to reach the statute books, even in England, where the first such laws were passed. In 1800 the first bill to outlaw bull-baiting had ignominiously failed in its passage though the Houses of Parliament, opposed by George Canning (later prime minister), who claimed that bull-baiting "inspired courage and produced a nobleness of sentiment and elevation of mind. . . . Putting a stop to bull-baiting was legislating against the spirit and genius of almost every country and age." However, in 1821 Dick Martin, MP for Galway, forced through a bill to protect horses and cattle against ill treatment. It was the first law of its kind in any country. In 1824 the Society for the Prevention of Cruelty to Animals was formed at the unfortunately named Old Slaughter Coffee House in London. The publication in 1859 of
Darwin's[4] *Origin of Species* seemed to strengthen the relationship 4 72 *143*
between humans and animals and support the animal-defense argu- 4 74 *145*
ment. In 1876 the Victoria Street Society against Vivisection was formed with Lord Shaftesbury as chairman. The same year a bill was

passed to prevent the vivisection of dogs, cats, mules, horses and asses. By the late nineteenth century the animal-defense movement had spread throughout the Western world and given birth to hundreds of local groups known as Humane Societies, in spite of the fact that the name more properly belonged to earlier humanitarian work of an entirely different nature.

5 133 *251* The Royal Humane Society[5] was founded in London in 1774 largely as the result of the efforts of Dr. William Hawes to promote knowledge of artificial-respiration techniques. Hawes based his ideas on the translation of a paper by the Amsterdam Society for the Recovery of the Apparently Drowned. The society had been founded in 1767 after several cases of successful resuscitation had been reported in Switzerland. In the nineteenth century interest in drowning became acute with the spectacular increase in cargo tonnage and passenger traffic on the high seas following the spread of industrialization. As the number of ships rose so did the number of shipwrecks and deaths.

From time to time the Royal Humane Society awarded a gold medallion for outstanding feats of bravery, and in 1838 the recipient hit the front pages because she was a slightly built, twenty-two-year-old woman. On the night of February 6, a paddle steamer, the *Forfarshire,* battling through a gale en route from Hull to Dundee with a full cargo and sixty-three passengers, sprang a leak in her boiler. The captain decided to take shelter among the Farne Islands off the coast of Northumberland. During this maneuver the ship hit the rocks and broke in two, and all but thirteen passengers and crew were drowned. The survivors, exposed to the full force of the storm, included a mother and two children. Overnight the two children and an adult died. At five o'clock the next morning Grace Darling, daughter of the local lighthouse keeper, caught sight of the wreck and the survivors clinging to the rocks. Grace and her father rowed to the rescue, struggling through mountainous sea in a small open boat. The drama was reported in the newspapers and Grace became an instant national hero. Alas, she was to die four years later from tuberculosis. Meantime she had inspired the public to offer massive fi-

Fig. 1: *Henry Greathead's "Original," in 1890 the first purpose-built, oars-only lifeboat designed without a rudder to be steered either way.*

nancial and political support for the eventual establishment of the Royal National Lifeboat Institution, in 1854.

That same year came another highly publicized loss at sea. The USS *San Francisco,* an American troopship carrying hundreds of soldiers, foundered in an Atlantic hurricane. The secretary of the navy
sent for Matthew Maury,[6] the only man in America who would be 6 31 *63*
able to tell where to look for survivors. After studying his wind and current charts, Maury pinpointed the spot and the survivors were found in the water.

Maury was the fourth son of a Huguenot-English family long settled in Virginia (his grandfather had taught Thomas Jefferson), and he had joined the U.S. Navy in 1825. It was during a voyage to South America that Maury became interested in finding faster ways to cross the ocean. On his return in 1834 he took leave and wrote his first work on navigation. In 1839 Maury published a series of articles in the *Southern Literary Messenger,* one of which advocated the establishment of a naval school. It would become the U.S. Naval Academy at Annapolis.

In 1847 Maury issued the first of several charts and then, in 1851, *Explanations and Sailing Directions to Accompany the Wind and Current Charts.* At the instigation of the U.S. government, copies of the charts and *Sailing Directions* were distributed free to all masters of vessels on the understanding that they would keep a full log of journeys and forward these logs to Maury, in Washington. Logs were to include temperature of air and water, direction of wind and currents, and air pressure. Captains were also required to throw overboard (at given intervals) a bottle containing a piece of paper carrying the ship's position and the date. They were also to pick up any such bottles they came across and note all details in their logs. In return for these services masters would receive free copies of Maury's further work. Over eight years Maury collected and processed data on many millions of observations, as a result of which he was able to identify faster sailing routes. One ship's master following Maury's suggested route from New York to Rio de Janeiro halved the usual journey time. It was reckoned that Maury's "Path-of-Minimum-Time" routes saved American shipping forty million dollars a year.

In 1853 Maury crowned his career when he persuaded sixteen
countries (among them the United States, Britain, Belgium, Holland,
7 28 *57* Russia, France, Norway, Denmark and Portugal) to meet in Brussels[7]
for the first International Meteorological Congress "to plan an uniform system of meteorological observation at sea, and to agree a plan for the observation of the winds and currents of the oceans with a view to improving navigation and to enrich our knowledge of the laws which govern those elements." Not long after he had returned from Brussels, Maury received a letter from a retired paper-
8 31 *63* manufacturing millionaire named Cyrus W. Field,[8] who was seeking
advice on the ideal route for a transatlantic submarine telegraph cable.

Submarine cables had already been laid successfully in the relatively shallow waters between England and Holland, Scotland and Ireland, but the Atlantic represented a formidable challenge. Field had managed to get a favorable charter from the British government for a fifty-year monopoly on any cable laid between Newfoundland and Ireland. The British also offered to provide a cable-laying ship as well as a generous advance on income from telegraph messages. Field then spent two years laying a cable between Newfoundland and the North American mainland (stockholders in the company included such luminaries as Lady Byron and Thackeray). When the link was completed Field wrote to Maury to solicit his views on the best route out of Newfoundland toward Europe.

Maury reported that soundings revealed a shallow "telegraph plateau" across much of the North Atlantic, and in 1857 work began on laying the cable. After a few hundred miles had been laid the cable snapped. Three more attempts were made and on August 5, 1858, 1,850 miles of copper wire connected Valencia, Ireland, with Trinity Bay, Newfoundland, and traffic began with an inaugural message from Queen Victoria to President Buchanan. At the celebration dinner in New York Field said modestly: "Maury furnished the brains, England gave the money, and I did the work." Then the cable failed again. In 1865 they found the parted ends, spliced them and the work was done. The U.S. Congress voted Field a gold medal.

Field had also written to the man whose work had inspired the
whole venture: Samuel Morse,[9] inventor of the most successful form **9** 32 *64*
of telegraph. Morse's advantages over other telegraphers were his key and the Morse Code, which he demonstrated before Congress in 1844. The idea had come to him in the autumn of 1832 during a voyage back to the United States from France. Morse first learned what he needed to know about the principles of electricity and one of his friends, Alfred Vail, provided the finance and hardware (Vail's father had a machine shop in New Jersey). Vail also suggested what would later become known as the Morse Code.

At this time Morse was a well-known artist, professor of art at New York University, and had just spent three years in Europe studying and painting. Morse was a strange man given to apocalyptic patriotic views. He had been brought up as a strict Calvinist by his father Jedidiah, America's foremost geography scholar, who had earlier led the Old Calvinist "Great Awakening" crusade against liberal theology. Like his father, Morse looked forward to the triumph of American culture and believed that only an elite could lead the country to salvation. Morse was also extremely xenophobic. At one point he painted a picture of the pope conspiring to arm American Catholics, provoke disorder, rig elections and elect foreigners to public office. Morse also helped publish a book about Maria Monk, a woman who claimed to have been a nun in Montreal, where she also claimed to have witnessed unnatural sexual acts performed by clergy and to have seen crypts filled with the corpses of illegitimate children. In the end it was revealed that Monk (rumored to have had a romantic affair with Morse) had escaped from a mental institution.

Morse believed that art was a tool placed in his hands by God to be used to save Protestant America. He believed that the millennium was imminent, and that when it came America would carry the empire of peace to the world. It was therefore essential to prepare American art for the great day. Morse founded the National Academy of Arts and Design in 1826 and was its president until 1845. The aim of the academy was to foster American artistic talent so that American genius could take its rightful place in the world and inculcate true Protestant virtues in other Americans.

In 1829 Morse decided to visit Europe to study artistic masterworks in preparation for what he hoped would be his greatest triumph, the commission to paint the four remaining murals for the Rotunda of the Capitol Building in Washington D.C. To this end, while in Paris in 1831, he painted the giant *Gallery of the Louvre*. The painting reproduced in miniature thirty-eight Louvre masterpieces. Morse's aim was to show that while the classical past was worthy of study it should not be the subject of slavish emulation by American artists, who, like the artist shown in the *Louvre* painting (Morse himself), could learn from the Old Masters and then develop their own distinctively American style. On his return the *Louvre* painting was put on exhibition in New York and was a disastrous flop. The commission for the Rotunda murals went to other artists. Morse turned to the telegraph as an alternative tool with which to make Protestant America great. Communications technology would be an instrument of Divine Will, redeeming America by transmitting messages of peace and love. At his demonstration in Congress in 1844 Morse's first transmitted message echoed these beliefs: "What hath God wrought!"

Morse had learned his art at the feet of Washington Allston, the most completely Romantic American painter, whom he had met in Boston in 1810 and with whom he became life-long friends. Only a year after their meeting Allston inspired Morse to attempt the first of his grand historical American scenes, *Landing of the Pilgrims at Plymouth.* That same year Morse joined Allston and his wife on his first trip to Europe. Allston was a good-looking Harvard-educated gentleman from South Carolina who on the death of his step-father in 1801 had sold the family property to finance a career in painting. On his earlier visit to London Allston had studied with Benjamin West, the president of the Royal Academy, and then in 1804 moved on via Paris to Rome. There he met Washington Irving, who later wrote: "I do not think I have ever been more completely captivated on a first acquaintance. He was of a light and graceful form, with large blue eyes, and black, silken hair waving and curling around a pale, expressive countenance. A young man's intimacy took place immediately between us, and we were much together during my brief sojourn at Rome. . . . We visited together some of the finest collections of paintings, and he taught me how to visit them to the most advantage, guiding me always to the masterpieces, and passing by the others without notice." Allston's *Italian Landscape* shows the profound effect of Italy on his work. His fresh New England eye was overwhelmed by the light, the color, the ancient ruins, the landscape dotted with hilltop villages, the rich mingling of Renaissance, medieval and classical architecture and the pastoral nature of Italian peasant life.

In 1805 Allston met and painted the English Romantic poet Samuel Taylor Coleridge, whom he would later recognize as his greatest intellectual mentor. At the time of their meeting Coleridge was suffering the aftereffects of his failure to give up opium. Aged thirty-three, with "Kubla Khan" and the "Ancient Mariner" poems behind him, Coleridge was already famous. He was also an alcoholic, deeply in debt, unhappily married with three children, had failed in a venture to set up a utopian settlement on the banks of the Susquehanna in Pennsylvania, and was extremely hypochondriac (he coined the word "psychosomatic").

It was partly to try to wean himself from opium (and his penchant for taking it in brandy), and partly to get away from his wife, that in 1804 Coleridge had run away to Malta. There, thanks to an influential acquaintance, he landed the job of secretary to the British civil commissioner, Alexander Ball. The post also included free food and accommodation in the commissioner's palace in Valetta, the island's capital. Coleridge's workload was light and consisted principally of rewriting Ball's dispatches to London. Although Coleridge complained incessantly about his health, the nightmare of withdrawal symptoms, the dull company and his inability to write new poems, he enjoyed the climate and the countryside and managed to produce some of his best prose. He also began to feel the first stirrings of mortality: "I had felt the Truth; but never saw it before clearly; it came upon me at Malta, under the melancholy dreadful feeling of finding myself to be Man, by a distinct division from Boyhood, Youth, 'Young Man.' Dreadful was the feeling—before that, life had flown on so that I had always been a Boy, as it were—and this sensation had blended in all my conduct." When his friends William and Mary Wordsworth saw him on his return to England, they were to remark that he had changed for the worse.

Coleridge saw from the dispatches he was editing that he had arrived in Malta at a critical time. Ball was arguing the strategic impor-
10 43 *78* tance of the island now that Napoleon[10] had given up Louisiana,[11]
10 59 *120* lost Santo Domingo and would inevitably turn his attention to the
10 112 *212* Mediterranean. Ball also suggested to the British government that Al-
11 67 *139* giers, Tunis and Tripoli were ripe for colonization and "they are capable of growing all our colonial produce." He also argued that though both Russia and France wanted Malta they should not be allowed to take it. At the time, the island was a hotbed of intrigue. The Maltese were agitating for independence, Russian and French spies were imagined to be everywhere and there was an American naval squadron on station commanded by Commodore Edward Preble.
12 61 *123* Among Preble's officers was the young Stephen Decatur,[12] leader of the daring and successful 1804 raid on Tripoli harbor to destroy the American frigate *Philadelphia,* which had run aground and been captured during the American-Tripolitanian War. During a brief trip to Sicily Coleridge met and dined with both intrepid Americans, and for years later regaled friends with tales of their exploits.

Coleridge's employer, Rear Admiral Alexander Ball, had joined the British Navy at the age of twelve. This event, he told Coleridge, had been inspired by reading *Robinson Crusoe*. Ball had the air more of an academic than a sailor, bookish and thoughtful. After serving in the Caribbean, America and Newfoundland, in 1783 he took a year off and went to France to study the language. At one point there, during a visit to St. Omer, he met another young captain with whom his fate was to be bound up, in spite of the fact that on this occasion each expected the other to make the required formal call, so neither did so. Ball then served in the English Channel, went again to Newfoundland, was stationed off the French Coast and in 1798 was posted to the Mediterranean where he was to meet the young captain with whom he had failed to exchange courtesies in St. Omer. At this time Britain was expecting to be invaded by Napoleon, and much of the British fleet was patrolling outside French harbors in the English Channel and on the French Atlantic coast. Hearing a rumor that Napoleon was assembling a Mediterranean fleet in Toulon, the British also sent a fleet to blockade that port.

In April the Toulon blockade fleet fleet was joined by a small squadron under the command of the man Ball had met in France, Captain (by now Admiral) Horatio Nelson, the fastest-rising star in the British Navy. No sooner had Nelson's squadron arrived off Toulon than it was blown south and scattered by a ferocious gale. Off the coast of Sardinia Nelson's flagship lost its mainmast as well as much of its rigging and was being driven by mountainous seas toward a rocky coast. At the last minute Ball brought his own ship alongside, took Nelson's flagship in tow in spite of Nelson's orders to the contrary and saved the day. The captain of the flagship later reported that Ball used his speaking trumpet to call "with great solemnity and without the least disturbance of temper . . . 'I feel confident that I can bring you in safe. I therefore must not, and by the help of Almighty God will not leave you.'" After the rescue Ball was mentioned in dispatches and the two men remained close friends for the rest of Nelson's brief life.

Meanwhile Napoleon's fleet had taken advantage of the situation to slip out of Toulon and head for Egypt. On the way there Napoleon sent a detachment to capture Malta. Ball was now dispatched (with Nelson's influential approval) to retake the island. After a two-year

Fig. 2: *The great British hero Admiral Lord Nelson, who was killed by a French sniper at the Battle of Trafalgar, aged forty-seven.*

siege he did so and was appointed commissioner. Nelson himself caught up with Napoleon at the Battle of the Nile, where he defeated the French, and then sailed back to Naples for repairs. In Naples Nelson would fall in love for the fifth time (after the daughter of Quebec's provost marshal, a clergyman's daughter, the wife of the commissioner of Antigua, and Fanny Nisbet, niece of the president of the St. Nevis Council). In 1787, Fanny had become Nelson's wife, to the chagrin of his colleagues, one of whom remarked that Fanny had two remarkable attributes: her good complexion and a "remarkable absence of intellectual endowment."

Ten years after the marriage the heroic Nelson arrived in Naples. He was Europe's unlikeliest heart-throb: he was thirty-eight, short, plump, white-haired, with a squeaky voice and Norfolk accent, blind in the right eye and with a stump for a left arm. He was in the habit of introducing himself: "I am Lord Nelson and this [gesturing to his good arm] is his fin."

The object of Nelson's Neapolitan infatuation (and later, while her husband was still alive, his mistress and later still mother of his two children) was a thirty-three-year-old married Englishwoman with a hidden past and the habit of wearing no underwear. Lady Emma Hamilton was the wife of Sir William Hamilton,[13] the sixty-seven- **13** 131 *249*
year-old English minister to the Court of Naples. In 1785 Hamilton had, so to speak, taken Emma over from his impecunious nephew Greville to clear the decks for the latter's impending marriage into a rich family. Emma was not told of the arrangement, merely that she would be living in Naples for six months until Greville could return for her. Nine months later, realizing that Greville was not coming, Emma gave in to the widower Hamilton and they became lovers.

To please Emma, Hamilton arranged singing and music lessons, trips to the newly excavated ruins of Pompeii and Herculaneum, rides up Vesuvius and a series of *conversazioni* at which Emma was introduced to the local nobility and the Neapolitan royal family. Soon she had become famous for her "entertainments," in which she posed, diapahanously dressed, in various classical tableaux: Agrippina scattering the ashes of Germanicus, Orestes sacrificing his sister, Oedipus blinded, and (very popular) the Bacchante "surprised while bathing." In 1791 Hamilton returned briefly to England to marry Emma. The couple then returned to Naples where Hamilton continued "collecting" antiques from archeological sites to sell in London, where his Greek and Roman vases would inspire the potter Josiah
Wedgwood and help kick off the Neoclassical[14] movement. **14** 80 *152*

In the light of Emma's past, her marriage to Hamilton made quite a furor. Emma had begun life as Emma Lyon, daughter of a humble smith. Brought up in Wales, at the age of twelve she was already employed as a nursemaid. A year later Emma was in London, maid to Mrs. Kelly, a well-known madame, and soon thereafter became one of Mrs. Kelly's "girls." By the age of sixteen she was living with a "protector," Harry Featherstonehaugh, who then passed her on to William Hamilton's nephew, Greville.

In London Emma was also rumored to have been employed as a "maiden" at Dr. James Graham's wildly fashionable Temple of Health, where patrons, including the duchess of Devonshire, were given electric shocks and served by transparently dressed "attendants." Graham's Temple was an elaborately decorated Adam house in The Adelphi, London. The fencing master to George IV later recalled "carriages drawing up next to the door of this modern Paphos, with crowds of gaping sparks on either side, to discover who were the visitors, but the ladies' faces were covered, all going incognito. At the door stood two gigantic porters, with each a long staff, with ornamental silver head, like those borne by parish beadles, and wearing superb liveries, with large, gold-laced cocked hats, each was near seven feet in height, and retained to keep the entrance clear." Entering under an enormous gold star, Graham's clients found themselves in lavishly decorated rooms with stained-glass windows. Music played and perfume drifted on the air. In these salubrious surroundings the elite were treated with medications including Nervous Aetherial Balsam, Electrical Aether and Imperial Pills. The star of the show had been made for Graham by an expert tinsmith. It was the Magnetico-Electrico Celestial Bed, on which childless couples would receive electrical shocks while coupling. The shocks were said to ensure immediate conception.

Graham may have derived his interest in electricity from conver-
15 134 *251* sations with Benjamin Franklin[15] in Paris in 1779 when he met
Franklin during the latter's time as U.S. minister to France. Electricity was the subject of much speculative experimentation at the time. In 1720 the Englishman Stephen Grey had electrified a small, suspended boy. In 1743 Johann Kruger, professor at Helmstadt University, had suggested that passing an "effluvium" of electricity through the body might be good for the health. Christian Ratzenstein claimed electric shocks raised the pulse rate and increased the circulation of the blood. Samuel Quellmaltz said he had used electricity to cure paralysis of the hand well enough for his patient to play the klavier. Even such respectable persons as John Wesley recommended electric treatment for nervous disorders. In 1777 an electrical machine was

ordered for St Bartholemew's Hospital in London. Graham is often described as a quack, but in an age when much medicine was still guesswork and mumbo-jumbo, he may have been no worse than anybody else. Besides, he had received his training at Edinburgh University, site of the best medical school in Britain, where he attended lectures by the great Joseph Black.

Black was a high-flier who had made his international reputation by the age of twenty-seven, when in 1755 he published a paper on an experiment in which he had heated limestone and caused it to become caustic quicklime. The value of this particular bit of research was that all contemporary treatments for kidney stone, common at the time, involved caustics. Up to then it had been thought that quicklime acquired its causticity from the fire. Black proved otherwise and changed the course of chemistry. He found that quicklime was made when a gas in the limestone was driven out by the heat, and that this gas could be recombined with quicklime to form limestone once again. Moreover, this combining and recombining process could be continued without limit. Each time the volume and weight of the relevant components were exactly the same.

The other world-changing discovery by Black came as a result of his investigation into the process of distillation. The Scottish whisky manufacturers' market had expanded rapidly since the union of Scotland with England in 1707, and distillers were keen to find ways of getting more whisky for less cost, so Black was concentrating his researches on finding ways to save fuel. His experiments on the amount of heat required to boil off liquids revealed the existence of latent heat, which explained the extremely high temperature of steam and why the distillers needed such copious quantities of cold water to condense it.

The latent heat discovery also showed James Watt[16] (who worked **16** 37 *68*
at the University of Glasgow when Black was teaching there) why the **16** 139 *259*
Newcomen steam-driven pump was so inefficient. In the pump (one of which Watt was repairing at the time) steam entered a cylinder kept ice-cold with a water jacket. This caused the steam entering the cylinder immediately to condense, creating a partial vacuum and al-

lowing air pressure to force the cylinder piston down. The piston rod was attached to one end of a pivoted beam set above the cylinder. As the piston moved down the other end of the beam moved up, lifting a rod attached to a suction pump. The problem was that the high temperature of the steam was heating the cylinder too much and weakening the subsequent condensing process on each stroke until the cylinder became so hot that no condensation would take place and the pumping action would stop. Black showed Watt that the cylinder would have to be linked to a separate, chilled condensing chamber (immersed in cold water) so that the scaldingly hot steam could condense there and not heat the cylinder while doing so. This separate condenser was the secret of Watt's success.

Through his association with Black in 1769 Watt was able to carry out some of his key experiments at Kinneil House, outside Edinburgh. This was the ducal seat of the Hamilton family and leased
17 36 *68* by Dr. John Roebuck,[17] a successful entrepreneur and ex-pupil of
Black's. Roebuck owned a coal mine that supplied his Carron ironworks, and since the coal mine was subject to flooding, his hope was that a successful Watt pump might save it. In the event the mine flooded early and drove Roebuck to bankruptcy, but not before he had helped finance Watt's research by paying off his debts in exchange for a percentage of Watt's patent rights on the steam pump. When bankruptcy intervened in 1772 Roebuck sold his share of
18 38 *68* Watt's patent to Matthew Boulton,[18] a shoe-buckle manufacturer in
18 97 *178* Birmingham, and Watt finally met the partner he needed. Together
Boulton and Watt turned the steam pump into the engine of the Industrial Revolution.

Meanwhile, Roebuck had already made his own contribution to industry with a new process for the manufacture of sulphuric acid. Not long after graduation he had invented an improvement in refining precious metals. This process used sulphuric acid, and in 1749 Roebuck set up a new sulphuric acid manufacturing facility at Prestonpans, near Edinburgh. The earlier manufacturing technique had involved burning sulphur and niter over water and condensing the acid from the fumes in glass globes. Roebuck replaced the glass globes with small lead chambers and quartered the manufacturing costs.

The market for sulphuric acid was growing steadily as the textile industry became more mechanized. By 1760 John Kay's flying shuttle was in general use, doubling the output of weft threads. Nearly ten years later James Hargreaves's spinning jenny multiplied the number of weft thread spindles that could be worked by one spinner. Richard Arkwright's 1769 water frame drew out the weft thread on mechanically rotating cylinders, and in 1779 Samuel Crompton's mule combined the jenny and the frame to complete the mechanization of the entire process. The mule produced thread fine enough for the best muslin cottons. By this time the market for cotton was booming, quadrupling raw cotton imports between 1791 and 1800.

The growth of cotton manufacture triggered a consequent rise in the demand for bleach. Before Roebuck's new system traditional bleaching of cloth (to rid it of its natural gray-yellow color) had been done in bleachfields. Between March and September the cloth to be bleached was stretched out, doused with fermented milk and left for six weeks to whiten. Roebuck's cheap, diluted sulphuric acid would do the same job in twenty-four hours. In 1785 the French chemist C. L. Berthollet had discovered that chlorine gas was a powerful bleaching agent, and James Watt introduced its use in Scotland. Cloth to be bleached was hung in gas-filled rooms, where bleachers sometimes died from the effects of breathing the gas. Then in 1799 Charles Tennant passed the gas over slaked lime and produced the first safe, cheap bleaching powder.

As an almost immediate result white paper became common. Before this, as can be seen from the gray tinge of English paper and the muddy color of early American documents, the color of paper depended on the rags from which it was made. Paper was made by pounding rags to a pulp, leaving them in water to ferment, pounding them again, then straining out the water on a vibrating wire mesh belt, and finally drying it by rolling the pulp between felt cloths and heated rollers. The first version of this process was invented by a Frenchman, Louis Robert, in 1799. The width of the paper made by his machine was that required for wallpaper, the fastest-growing furnishing accessory in Europe at the time. The Paris *Journal des Inventions* reported: "For the view, the cleanliness, the freshness and the elegance, these papers are preferable to the rich materials of the past; they do not allow any access to insects, and when they are varnished, they retain all the vivacity and charm of their colours for a long time. Finally, they can be changed very frequently . . . making us thus inclined to renovate our homes, cleaning them more often and making them gayer and more attractive."

When Robert's paper-making venture failed for lack of financial support in France he sold the patent to his erstwhile employer, Didot
19 113 *213* Leger. Didot's English brother-in-law John Gamble[19] then took it to Britain, where the paper-making brothers Fourdrinier set up the first fully operational version of Robert's process at their mill in Frogmore, near London, in 1808. In 1836, when the British government repealed the high wallpaper tax, mass production began. In 1839 Harold Potter of the Darwen wallpaper factory perfected a power-driven roller printer. By 1850 machines were able to print perfectly registered patterns in eight colors on fifty-four thousand feet of paper a day. The effect on the wallpaper industry between 1834 and 1860 was to increase output from a million to nine million yards. Prices dropped like a stone. What had been a luxury item was now available to all but the very poor.

An Englishman named William Morris used the new manufacturing and printing techniques to put wallpaper into the homes of the industrial middle class for the first time. Morris was a well-off Oxford graduate influenced like others of his age by the social conditions of industrial-age Britain. In the mid nineteenth century government surveys were beginning to reveal the scale of social problems that had been created by the rapid industrialization of the previous decades and exacerbated by overcrowded living conditions and the widening gulf between the rich factory owners and their disfranchised, poverty-stricken workers. Morris and his friends turned away from the horrors of the cities to medieval art and architecture. For them the Middle Ages represented an age of innocence when the craftsman had been an independent, creative spirit, free to take his skills wherever he chose, protected from exploitation by professional guilds.

Morris led the new Arts and Crafts movement dedicated to bringing this medieval sweetness and light to urban homes. From 1877 his company showrooms in Oxford Street displayed "traditional" furniture, tapestries and wallpaper with simple floral patterns based on medieval designs. The style revolutionized public taste. One social commentator wrote: "It may be questioned whether the decorative treatment of the walls should give place to pictures in rooms which are occupied from day to day. If we imagine the tired man of business returning to his suburban home . . . it can hardly be supposed that he will be in a position to make the special mental effort involved in inspecting his pictures; but supposing him to be the happy possessor of a harmoniously decorated room, he will at once be soothed and charmed by its very atmosphere."

Morris took his artistic views into his political life. The utopian spirit of his art was mirrored by socialist beliefs that drove him to what he called "holy warfare" against capitalism. From 1877 he gave a series of lectures to working men in which he attacked the values of Victorian society. In 1883 he joined the Democratic Federation, inspired by Marx's views on the alienation of industrial workers. Morris sold the federation's weekly paper, *Justice,* on the streets, and joined its executive committee together with Marx's daughter, Eleanor. In 1884 he broke with the federation (when it proposed to become an orthodox political party) and set up the Socialist League. When this in turn was infiltrated by anarchists he broke away again and founded his own Socialist Society in Hammersmith, London.

At the Society's musical evenings, where socialist songs were sung under the direction of composer Gustav Holst, two society members played piano duets. One of them was a twenty-seven-year-old would-be journalist named George Bernard Shaw, his ragged cuffs trimmed with scissors, wearing shabby, cracked boots and an ancient coat and

Fig. 3: *Annie Besant, free-thinker, social reformer and hygienist, who played piano duets with George Bernard Shaw.*

sporting a red beard and an Irish brogue. The other, one of Morris's ex-colleagues on the Democratic Federation executive, was the equally charismatic Annie Besant.

Besant was by this time a well-known activist, having been involved fifteen years earlier in one of the most widely publicized trials of the nineteenth century. In 1877 she and the national Secular Society's Charles Bradlaugh had been sentenced to six months in prison and a fine for republishing a forty-year-old pamphlet by an American au-
20 35 67 thor, Charles Knowlton,[20] titled "Fruits of Philosophy." The pamphlet contained detailed instructions on contraception for young married couples. During the trial (for obscene publication), Besant and Bradlaugh spoke out eloquently about the new threat of over-

population that would result from slowly improving living conditions and sanitation; on the overcrowded conditions in slums, rife with immorality and incest; on the effects of pauperism that led to a death rate of one in three infants; and on the need for freedom from prosecution for those publicizing the facts about contraception. The sentence passed on Besant and Bradlaugh was quashed by the judge before the two defendants had left the dock. Besant had become the first woman ever to speak out publicly to advocate contraception and get away with it.

In 1889 Besant became a Theosophist, and by 1891 she was effectively running the Theosophist Society. Theosophists rejected material things, encouraged vegetarianism, sought to bring the universal brotherhood of all races, investigated latent psychic powers and studied ancient and modern religions and philosophies. In pursuit of all these aims, in 1893 Besant visited India and set up the Central Hindu College for the Study of Comparative Religion in Benares.
Earlier that same year she had visited the Chicago Exposition[21] and **21** 24 *49*
together with Indian Theosophists had participated in meetings of the "Parliament of Religions." They took the meetings by storm, with four thousand attendees turning up at the last session to hear them. The Theosophists' vegetarian message was greeted sympathetically in the United States. The first vegetarian society had been formed in Philadelphia in 1850. In 1858 Dr. Caleb Jackson founded a health center at Danville, New York, based on vegetarian principles and including cold water treatments.

In 1865 Danville was visited by a Seventh Day Adventist named Ellen White. Two years earlier she had had a vision in which she was told to eat only two meals a day, avoid meats, cake, lard or spices, and consume only bread, fruits, vegetables and water. In Danville Mrs. White had a further vision. This time the message she received was to set up another Danville. A year later she and her Adventist colleagues bought a farm on seven acres of land just outside a small Michigan town and opened their health center.

The center's rules were strict: no levity, no playing checkers, lots of oatmeal pudding and religion and cold water cures. The diet excluded tea and tobacco. Soon after the center opened it was in financial difficulties. The Adventists looked around for a new superintendent. They chose a young man who lived in the same Michigan town and had at the age of fourteen begun typesetting for the Adventist printing house. The church elders sponsored him through a course at Bellevue Medical College in New York, and in 1875 he graduated and took over what was now known as the Western Health Reform Institute. The first thing he did was to rename it the Medical and Surgical Sanitarium. The new superintendent had a natural eye for publicity. He dressed entirely in white, seemed to need no sleep, and was seen frequently with a cockatoo perched on his shoulder. He fostered vegetarian diets and set up the Three-Quarter-of-a-Century Club (healthy eating promoted longevity) and, in 1914, the Race Betterment Foundation. At the sanitarium he developed courses in nursing, physical education and home economics. For the patients he introduced room service, a gymnasium, a string orchestra in the dining room and wheelchair social events on the front lawn.

There was one aspect of the diet that still troubled him: "the half-cooked, pasty, dyspepsia-producing breakfast mush." To this end he began experiments in the sanitarium kitchen. Some time in 1894 he boiled and steamed wheat into a paste, flattened it between rollers, scraped the pieces emerging from the rollers and baked them crisp. In March 1895 at the General Conference of Adventists he presented his invention. It changed life in the Adventist community and then the world. The Superintendent of the Battle Creek Sanitarium was named John H. Kellogg, and his invention was named "cornflakes."

CHAPTER 2

What's in a Name?

There used to be as many different kinds of breakfast as there were different cultures to eat them. The few still served around the world include corn pancakes and curry (India), blood pudding and fried potatoes (England), waffles and maple syrup (America), cold ham and cheese (Germany) and meat-and-cabbage soup (Colombia). However, these quaint local anachronisms are gradually disappearing before the onslaught of television advertising campaigns that exhort viewers to "eat healthy." The breakfast-food marketplace is now global. Today in almost any store anywhere in the world you can buy breakfast cereals.

For every hundred kilograms of grain used to make cornflakes, eighteen kilograms of used corncob are left. This vegetable detritus has had a varied career. Early in the twentieth century it was used to increase the water-holding capacity of mulch and soil, as well as for landfill in swampy ground. Corncobs have also been used for animal and poultry feed, as poultry litter, as a mild abrasive to clean car windscreens, and in soft-grit blast-cleaning of metals in aero engines.

However, it was the work of a nineteenth-century German chemist named Wolfgang Dobreiner that turned corncobs into a world-changing product. Dobreiner rose from obscure beginnings to become an unqualified journeyman chemical manufacturer, and at the 22 69 *141*
age of thirty had the good fortune to befriend Johann Wolfgang von 22 71 *142*
Goethe,[22] chief administrator of the technical college in the Univer- 22 82 *153*

sity of Jena, Germany. The college patron, Grand Duke Carl August of Saxony-Weimar, probably approved Dobreiner's hire in the hope that he would produce potentially profit-making inventions. Whatever the reason, in 1810 the unqualified Dobreiner was given a doctorate and a place on the faculty. Twenty-two years later he found a use for corncobs.

Dobreiner processed corncobs (it is not known how, or why) and produced an amber-colored chemical he named "furfurol." Little or no use was made of this discovery until the 1920s, when the growing petroleum industry had begun to make inroads into the chemicals market. Up to this point most chemicals had been derived from plants, so the change was bad news for the agricultural industry. Quaker Oats looked around for more ways to make money from their products and found that pressing, boiling, steaming and acidifying oat husks (and other forms of bagasse, such as corncobs) would yield the almost-forgotten furfurol. It was then discovered that furfurol could be processed to make a solvent for use in oil refining, in the manufacture of synthetic rubber and the development of nylon. It also found uses in carbuncle ointment, antibacterial medicine, acid-resistant containers, molds for the metal industry, insecticides, charcoal for barbecues, herbicides and antiseptics.

Furfurol was also used as an adhesive resin that would bond abrasives to a grinding wheel. Up to the end of the nineteenth century the abrasives (emery or sandstone) wore out quickly. Then in 1891 a young American, Edward Goodrich Acheson, made an accidental discovery that changed grinding and illumination. Acheson had previously worked as timekeeper, railroad ticket agent, assistant surveyor and railroad engineer and oil-tank gauger. In 1880 he was inspired by an article in *Scientific American* to seek employment with
23 75 *147* Thomas Edison[23] at Menlo Park, where for four years he worked on electric lamps. In 1888 he set up his own small electric plant in Monongahela, Pennsylvania. Three years later he was using an electric-arc furnace to pass an extremely powerful electric current

through a mixture of clay and powdered coke (he may have been seeking to make artificial diamonds), when he noticed a few bright specks in the fused mass. Putting one speck on the end of a pencil-lead he drew it across a window pane and saw that it cut the glass. Using a furfurol derivative Acheson stuck many of these abrasive specks on a small wheel, took it to New York and sold it to a diamond cutter. Fliers posted to twelve thousand dentists brought enough response to finance a booth at the 1893 Chicago Exposition, and it was there that Acheson's grinding wheel came to enlighten the general public.

The Chicago Exposition[24] was the biggest of its kind ever staged. Be- **24** 21 *45*
tween May and October of that year over twenty-one million visitors would attend. Tenders had been requested for the Exposition's illumination, and General Electric put in a bid for lights at $13.98 each. There was also a bid from Charles F. Locksteadt of the Chicago South Side Machine and Metal Works for $5.25 a light. Locksteadt got the
contract and approached Charles Westinghouse[25] with a request for **25** 76 *147*
250,000 lamps. Since Edison owned the patent to the incandescent light bulb, Westinghouse designed a new type of bulb. Edison's patent was for a one-piece bulb, so Westinghouse designed a bulb that came in two pieces: the glass bulb and a separate, airtight glass plug, fitting into the end of the bulb and carrying the power leads and filament. The plug was airtight because it was in the form of a ground-glass stopper. The 250,000 glass stoppers for the Exposition were ground by sixty thousand small Acheson abrasive wheels.

Acheson called his new abrasive "carborundum." Today it is better known by its chemical name: silicon carbide. The material (and its sister product boron carbide) came into international prominence following the development of armor-piercing shells and bullets. Both carbides are the hardest-known ceramics and the hardest materials in existence after diamond. When either of the carbides is used as armor, when a bullet strikes it a cone-shaped depression is formed in the carbide. This is then pushed by the bullet into the softer, backing material of the armor. The area of the cone is wider than the cross-section of the bullet so the energy of impact is absorbed by the armor over a much bigger area, attenuating the force. At the same time the carbide pulverizes the bullet and spreads the impact energy even further. These characteristics made body armor popular when it reached its most refined stage of development in the 1960s Vietnam War, where it was used to protect Air Cavalry helicopter crews and ground forces. Aircrew armor reduced wounds 27 percent and fatalities 53 percent. On the ground, in hand-to-hand combat troops could throw hand grenades as close as thirty feet and absorb in their flak jackets the grenade blast that killed the enemy.

The need for armor-piercing weapons first emerged when navies began to switch from wooden ships to ironclads. In the 1860s, when the first of the new French ironclads was launched to the accompani-
26 117 *220* ment of belligerent noises from the French Emperor, Napoleon III,[26]
26 127 *244* the British responded by building their own. In the first ironclad-to-
26 129 *246* ironclad encounters cast-iron cannonballs fired against two feet of
wrought-iron cladding backed by a foot or more of solid teak proved ineffectual. It was a Captain Palliser of the British Eighteenth Hussars who first came up with the idea of a pointed shell with a hard nose. During the Chile-Peru war of 1879 a Palliser shell, aimed at the Peruvian ship *Huescar,* penetrated five and one-half inches of wrought-iron armor, then thirteen inches of teak and finally a half-inch of steel.

In the 1880s an American researcher made a major discovery. Charles E. Munroe had been working on gun cotton and found that if he exploded a slab of it against a slab of steel the words "U.S. Navy" incised on the gun cotton slab were reproduced by the blast in the surface of the steel. Early in World War I a German experimenter, J. Neuman, found that if the incision were lined with metal the impression on the steel was greatly deepened. In its final form the "Munroe effect" was achieved with ammunition holding a charge of high explosive in a cavity lined with copper plate. When the explosive was detonated at the end farthest from the cavity it collapsed the liner and created a focused blast in the form of a fine jet of hot gas and molten metal that punched through steel with ease.

World War I gave added impetus to the search for effective armor-piercing systems because of an entirely new battlefield development. It was an invention so secret that during the early stages of its manufacture by the British it was officially described as a water cistern being made for Russian factories. Because of this it became known as a "tank." The tank was initially developed because of two other inventions that had recently appeared on the battlefield: barbed wire and the machine gun. Because of the large number of casualties caused by machine guns when infantry were entangled by barbed wire, it was essential to find a way of breaking rapidly through the wire. The tank was developed to fulfill this function.

On November 20, 1918, at the battle of Cambrai, the first use of the tank changed the face of war. Behind a rolling artillery barrage 358 Allied tanks crept forward toward German positions heavily fortified with barbed wire. The tanks attacked in what became known as the "unicorn" formation: groups of three tanks, one central and leading the other two flanking it on either side, each protecting infantry advancing behind it. The tanks could also carry fascines (bundles of brushwood), which were dropped to fill trenches. The Cambrai at-

Fig. 4: *The terror weapon of World War I. Note the machine gun set in the swiveling sponson at the side.*

tack was a total success. Along a front stretching thirteen thousand yards, protected by the tanks the infantry advanced ten thousand yards in ten hours. Eight thousand prisoners and one hundred artillery pieces were captured. Allied casualties were minimal. The future of the tank was assured.

Part of the reason for the tank's success was its versatility. Mounted on two caterpillar tracks it could cross ditches and plowed ground or climb over banks, walls, hedges and fences. It could knock over small trees, cross streams over a foot deep, climb a slippery slope, handle a six-foot wall and survive a sheer drop of fifteen feet. It would do all this at a speed of three miles per hour. By the end of the war, the Mark V tank speed had increased to five miles per hour and it could turn in seventy-five feet. Only fifty of the Mark Vs were ready, but they made all the difference. In 1918 fifty-nine British divisions defeated ninety-nine German divisions because the British had the tanks and the Germans only had what had become known in German as "tank terror."

When the first tank (codenamed "Mother") was produced in 1915 it owed its existence to a friend of Major General Sir Earnest Swinton who had written to him about the new American tractor the British army was already using to haul supplies. A total of twelve hundred of these machines had already been bought by the Allies. They were tracked vehicles that would traverse almost any ground and had originally been developed near Stockton, in the San Joaquin Valley of California, where the rich, damp bottom land (which would not support a horse) convinced lumber merchants Benjamin and Charles Holt of the need for a haulage vehicle capable of working in such conditions. The advantage of a tracked vehicle was that whereas in a conventional vehicle with a wheel diameter of six feet four inches. the area of each wheel's contact with the soil (on which the weight of the vehicle rests) is only twenty-three inches, a track spreads the weight over an area more than three times larger. On November 24, 1904, while the Holts were testing their first steam-driven "crawler," a photographer friend, Charles Clements, remarked that it looked like a caterpillar. In 1910, when the firm was incorporated in Illinois, its name was the Holt Caterpillar Company.

The version sold to the Allies in Europe was gasoline-powered and was relatively maneuverable, with each track driven independently. In 1917 a new 120-horsepower version was produced. Then in 1925 came the first field trials of a new four-cylinder version with a radically different engine. This was the brainchild of a German who had already sold his engines to eleven countries. In Russia they were being used in power stations. In France they powered canal barges and submarines. The British installed them in battleships, the Dutch in passenger ships and the Germans in locomotives. The new engine had numerous advantages: it was more fuel-efficient, it was compact, it started cold, and it used cheap fuel. The engine's inventor was named Rudolf Diesel.

Born in 1858 the son of a bookbinder in Paris, Diesel attended the
Munich Polytechnic where he was taught by Carl von Linde,[27] the **27** 55 *115*
inventor of the refrigerator. Graduating with the best exam results ever, Diesel then worked for the Sulzer refrigerator factory in Winterthur, Switzerland, and then sold the fridges, first in Paris and then in Berlin. By 1890 Diesel was well known as an inventor with an obsession: to replace the steam engine. His aim was to make an engine thermodynamically efficient and capable of using a wide range of fuels. In February 1892 he filed the patent for his new engine (which he had wanted to call "Delta" or "Beta" but finally, at his wife's insistence, named after himself). The diesel engine was an internal combustion engine in which the fuel was injected into the cylinder where the air had been compressed so that its temperature rose to eight hundred degrees Centigrade, the fuel's spontaneous ignition point. Since there was no spark that might ignite the fuel at the wrong time in the piston cycle, the system was highly efficient.

The main attraction of the engine for Europeans was that it could run on fuels other than gasoline. In the early part of the twentieth century few European countries had their own oil supplies, and the cost of the fuel was considerable. Diesel engines would run on liquid fuels of many types: whale oil, tallow, paraffin oil, shale oil, naphtha, even peanut oil. Diesel mentioned the most attractive fuel of all in a speech at Kassel in 1897, when he said that the engine would reach its real potential only when it used common hard coal. This may have been the reason Diesel won a contract with Heinrich Buz's Augsburg engine factory. Buz's partner in the deal with Diesel was the coal king of Germany: Alfred Krupp.

The Krupp family had come to Essen in the sixteenth century, married into a local gun-making family, and became gunsmiths themselves. For the following three hundred years they were active in trade and in public office. Unlike all the other great German mercantile families, no Krupp ever took up a profession. In 1811 Friedrich Krupp left the spice trade and founded an iron and steel works in Essen. The city was ideally situated for foundries, located as it was among over a hundred local coal mines. In 1826 Friedrich's son Alfred (in charge of the company at the time of the Diesel deal) took over at the age of fourteen after his father had left family finances near rock bottom. For twenty years Alfred endured "perpetual grind and near-gloom" to make a success of the enterprise. He was aided by the 1834 German Customs Union, which created a single market of twenty-five million people. In 1851 Alfred exhibited a steel six-pounder gun at the Crystal Palace Exhibition in London and caused a sensation. At the same exhibition Krupp also showed the world's biggest steel casing, which weighed a staggering forty-three hundred pounds. The Exhibition made Krupp famous overnight and, more important, attracted the attention of Prince William, who later would become Kaiser and decorate Alfred with the Order of the Red Eagle with Oak Leaves, an honor usually reserved for victorious Prussian generals. Krupp went on to build the core of the new German navy: nine battleships, five light cruisers, thirty-three destroyers and ten submarines. By the early 1860s Krupp was already Europe's biggest cast-steel manufacturer, with representatives in every major city. He had built a fully integrated business that included iron-ore mines, collieries, iron- and steel-making foundries and railroads.

By 1890 Alfred was employing seventy thousand workers and facing the social problems that went with a large workforce in a country going through political turbulence. Alfred was a first-class organizer, saying once: "As pants the hart for cooling streams, so do I for regulation." He held his company together with the introduction of specially designed uniforms for the workers to wear at home, a sick fund, a burial fund and a pension fund. He built company housing, hostels, schools, hospitals, canteens, stores, bars, skittle alleys, baths, a church and a cemetery. Alfred was virulently anti-Marxist and regarded his unprecedented welfare schemes as a way of countering the revolutionary message of socialism. On one occasion, when some workers attempted to form cooperative food shops, he bought the shops and incorporated them into his own company stores, saying: "We must make certain that every worker's immediate thought is of the firm, and the interests of the factory, and that he is not tempted to mull over speculations in coffee, tobacco, sugar and raisins." The Krupp empire was a state within a state. Workers joined for life.

The Krupp technique for dealing with political radicalism inspired the most influential man in Germany at the time, Otto von Bismarck, who became prime minister of Prussia in 1862. Two years later both men met in Essen when Bismarck was returning to Berlin after negotiations in Paris. They discovered a mutual attraction to horses, guns, trees and misanthropy (Bismarck once wrote to his wife: "I have more to tell myself when I am with trees than when I am with men"). Both men suffered from megalomania and both were tyrannical and unscrupulous. Bismarck was concerned with preserving the Junker class against the revolutionaries, and he believed that guns were the final argument of kings. The two men were made for each other.

In 1883 Bismarck gave Germany his version of the Krupp welfare program with a National Health Insurance Scheme, which provided medical treatment and up to thirteen weeks' sick pay for three million low-paid workers and their families. Employees paid two-thirds of each premium and their employers paid the rest. A worker who was permanently disabled or sick for longer than thirteen weeks was given protection by the Accident Insurance Bill, which followed in 1884. In 1889 Bismarck introduced the world's first state pension for retired persons over seventy, and disablement pensions for adults of any age. As he said, the legislation was aimed at neutralizing the effect of socialism: "Whoever has a pension assured to him for his old age is more contented and easier to manage than a man who has none," and, "We must carry out what seems justified in the socialist

Fig. 5: *The Iron Chancellor, Otto von Bismarck. For him great national issues were to be solved with "blood and iron."*

programme and can be realised within the present framework of state and society." At the same time Bismarck curbed the Socialists with draconian legislation that outlawed strikes and jailed party activists.

To aid him with the introduction and administration of these major acts of social policy Bismarck ordered comprehensive data to be collected on the population at large. Over the previous two centuries growth in trade and industrialization had generated an increasing number of population surveys. In the seventeenth century data was often arrived at by crude methods, e.g. estimating the age and sex distribution of a community by multiplying an assumed average family size by the number of chimney pots. Another method was to analyze data on births and deaths found in the baptismal and burial records of parish churches. By Bismarck's time statistics had introduced a measure of certainty into the analysis of the data. Because of their potential military value to an enemy population figures were now a matter of national security.

The entire study of social statistics was to be changed by a man who profoundly influenced Bismarck's chief of statistics, Ernst Engel. This was the Belgian astronomer Adolphe Quetelet. Quetelet was born in Ghent in 1796 to a family of modest means. On the early death of his father he became a math teacher in a local school. By the time he was twenty-four his remarkable talents had gained him the chair of mathematics at the Brussels Athenaeum and membership of the Belgian Royal Academy of Science. Over the next fifty years he was to dominate the country's science. In 1820 Quetelet led the movement to build an observatory in Brussels and traveled widely so as to learn all he could about astronomy. By 1842 the observatory was in place and he became its resident astronomer for the next forty-two years. Meanwhile, he had already carried out observations on the regularity of meteor showers, sunspots and tides and begun the collection of hourly meteorological observations. It was this work that brought the first international meteorological conference to Brussels[28] in 1853. **28** 7 *30*

Quetelet extended his observations to periodicity in a wide variety of phenomena, including daily mean temperatures, times of foliation and leaf fall and the blooming of flowers (he noted that the common lilac flowers when the sum of the squares of the mean daily temperatures, counted from the last frost of the previous winter, adds up to 4,262 degrees Centigrade). Quetelet's ambition was to discover the order that lay behind the apparently random natural world.

In all this he was informed by his experiences in astronomy, where techniques of measurement and observation were perhaps more developed than in any other science. One particular mathematical tool used by astronomers was extremely valuable when dealing with data that had been derived from separate observations of a heavenly object and that revealed apparent irregularities in its behavior. The "law of least squares" (a technique for smoothing out extremes of difference in data) was originally applied in astronomy to calculate the most likely position of a planet or meteor whose movement had been observed too infrequently to provide continuous data. Using the least-squares technique Friedrich Gauss had been able to predict where a newly discovered asteroid, Ceres, would be found when only three measurements of its movement had been made before observers lost sight of it.

In 1826 Quetelet became regional correspondent for Belgium's statistical bureau and began the work for which he is now best remembered. His greatest contribution to statistics was the concept of the "average person." Quetelet believed that if such a person could be reproduced mathematically then a new science of social physics would reveal the natural laws governing human behavior, making possible the identification of deviation from the norm. This would remove the element of guesswork from all social planning. "Chance," said Quetelet, "that mysterious, much abused word, should be considered only a veil for our ignorance." In 1831 he published *The Growth of Man*, a statistical survey of human physical data, and *Criminal Tendencies*, a study of the individual's propensity to commit crime. In 1835 came Quetelet's greatest work: *Social Physics: Man and the De-*
29 47 *90* *velopment of His Faculties.*[29] The book contained a rigorous analysis of almost every aspect of human behavior, and it effectively founded modern social science.

In 1832 Quetelet was invited to Cambridge, England, to attend the third meeting of the new British Association for the Advancement of Science. At a meeting of a small group of scientists and mathematicians interested in statistics, discussion of Quetelet's work on suicide and crime led to the proposition that the BAAS should set up a statistics section. Not everybody in the association shared the enthusiasm for such an idea. The association president, Adam Sedgwick, warned in his concluding address that the new statistics section would have to follow strict rules: "For if we transgress our proper boundaries, go into provinces not belonging to us, and open a door of communication to the dreary world of politics, that instant will the foul Demon of discord find his way into our Eden of philosophy."

In spite of this and other objections, the precursor of the Royal Statistical Society was founded in 1834. Seldom can a new science have come at a more opportune moment. Following on the heels of the Industrial Revolution, rapid urbanization had brought hundreds of thousands of workers into the city factories. Their living conditions were unspeakably filthy and degrading. Thousands lived in tenement blocks surrounding courtyards filled with sewage. Families often found themselves living in cellars ankle-deep in water, crammed ten to a room. Prostitution and incest were the inevitable result. By the 1830s social discontent was rife. The middle classes were becoming alarmed at the possibility of revolution and turned for help to statistics.

Surveys were carried out less with the the aim of improving the lot of factory workers and their families than of finding out the causes of the moral decay that rendered them disobedient to authority. Mathematics might provide the means with which to control the masses. To this end attempts were made to discover how many women could knit, or sing a jolly song, how many owned books and were literate, how many had insurance, or hung improving prints on their walls. Surveys sought to discover the workers' religious persuasions, the number of "amatory" pictures in their possession, how many of them cultivated flowers and how often they had their hair cut.

Hairdressers changed the life of the president of the new British Statistical Society, one of the members of Quetelet's Cambridge discussion group. He was Charles Babbage, at this time already one of the most famous figures in British science. Even for Victorian times Babbage was an extraordinary polymath. He was an inventor, mathematician, philosopher, scientist, outspoken critic of the scientific establishment, raconteur, political economist, socialite, visionary and prolific writer. He designed a recorder for monitoring the condition of railway tracks, lights for communication with ships, an ophthalmoscope, a pen with rotating discs for drawing dotted lines on maps, a system for delivering messages via aerial cables and footwear for walking on water. He also suggested designs for a tugboat, as well as for submarines propelled by compressed air, diving bells, an altimeter, a seismograph, a hydrofoil, a coronagraph to make artificial eclipses, a release-coupling for railway carriages, speaking tubes to link London and Liverpool and two kinds of cow-catcher for use on locomotives.

In 1819, during a visit to France, Babbage heard that the Baron De Prony had been commissioned to produce a new set of logarithmic and trigonometric tables for the recently introduced metric system. The Baron had brought together a group of "computers" to do the thousands of additions and subtractions required for the tables. In the new Republic complicated aristocratic hair styles were no longer fashionable, so many of these "computers" were out-of-work hairdressers. Babbage was stuck by the "division of mental labor" manifested in this French exercise, and this may have given him the idea of trying to automate such work. With the growth of commerce new surveys and tables were being produced every day, and they were subject to human error, which could cost money. In 1834 the science writer Dionysius Lardner wrote that a random selection of forty volumes of numerical tables revealed no fewer than thirty-seven hundred errors. Babbage himself reckoned that mistakes of this kind were costing the government up to 3 million pounds a year.

In 1834 his answer to the problem was to build an automatic calculating machine that would add and subtract. He also designed a more advanced, multiplying-and-dividing version (the "Analytical Engine"), which would foreshadow the modern computer. The core of this machine was a set of rods on each of which was mounted a series of independently rotating toothed wheels. Each wheel, with the numbers zero to nine marked on its edge, represented a numerical decade. The wheels interacted through complex sets of interlocking rods and cams, to produce addition, subtraction, multiplication and division. Sums were indicated by the number appearing in a small window set alongside each wheel in the outer casing. The Analytical Engine was also capable of carrying out complex calculations by means of stored programs. Addition or subtraction took a few seconds, multiplication and division a few minutes.

The operation was controlled by punched cards. A constant could be introduced into the machine's calculations by a "number-card." A "variable-card" defined the rod on which a number was to be placed or entered into the "mill" (the unit holding the stored programs for addition, subtraction, multiplication and division). A third type of card, the "operation-card," controlled which stored program was to

be used. Babbage's use of the cards solicited a poetic description from
his backer and colleague Lady Ada Lovelace (Byron's[30] daughter): **30** 60 *122*
"We may see most aptly that the Analytical Engine weaves algebraic **30** 132 *249*
patterns just as the Jacquard loom weaves flowers and leaves."

Lady Lovelace was referring to the punched-card system originally developed for the French-designed Jacquard silk-loom, in which sprung hooks were pressed up against one of a series of cards held on a belt above the loom. When a particular thread or set of threads was required to be lifted during weaving the relevant control card presented a set of holes through which the necessary hooks could pass and lift the threads, thus automating the patterning process. In the late nineteenth century the same cards would be used (with electrified wires instead of hooks) to automate the American census. The engineer who developed the system, Herman Hollerith, set up a business that would eventually become known as IBM.

Meanwhile, in Babbage's time others were using the same cards for very different purposes. In 1844 the British Parliament decided to open a rail link to run through Wales to the port of Holyhead on the Irish Sea. This meant crossing the Menai Straits, a rocky channel about three quarters of a mile wide. The Admiralty insisted that its tall-masted ships be able to go under the bridge, and on these grounds vetoed the use of a cast-iron arch. Several suspension bridges had recently fallen down and were considered unsuitable. The contract went to Robert Stephenson, a locomotive engineer and bridge-builder who had spent most of his life working on the railroads, and he decided on a revolutionary new scheme.

His fifteen-hundred-foot-long Britannia Bridge over the straits consisted of two gigantic wrought-iron tubes, each made up of four shorter tubes riveted together. The tubes would each carry one railroad trackbed and be large enough for a train to pass through. The tubes would be supported by three masonry towers and held on an abutment at each end. Since the longest wrought-iron bridge span built to this date was thirty-one feet, Britannia was totally unprecedented in both design and scale. The other new record to be set by the bridge was the number of rivets it required: 2,190,000. Such an amount of riveting could not possibly be completed by hand in time, and it fell to a Welsh tool-maker named Richard Roberts to solve the problem. Roberts produced an automatic, card-controlled device that would gang-punch holes through a wrought-iron plate as it passed through the machine. The punching machines could be thrown out of gear when not required. Those needed for a particular pattern of holes would be engaged via control rods operated by punched cards.

One of Stephenson's friends was another engineer, Isambard Kingdom Brunel, who was on site when the great iron tubes were floated into position below the bridge. He knew that all the tube stress-testing had been done by William Fairbairn, a prominent ironmaster and shipbuilder, who said at one point: "Provided we regard a vessel simply as a huge hollow beam or girder, we shall then be able to apply with approximate truth the simple formulae used in computing the strengths of the Britannia . . . and other tubular bridges."

These words were music to the ears of Brunel, who was about to begin work on the largest ship ever built, the *Great Eastern*. The idea for the ship was born of the Australian gold strike of 1851, which boosted emigration from Britain and added to the general growth in commerce between the two countries. Sailing ships completed the voyage in up to four months but were at the mercy of the winds. Australia was beyond the range of existing steamships because of the absence of sufficient bunkering facilities en route. A ship capable of doing the journey nonstop at an average speed of fourteen knots would take seventy days round trip and burn an average of 182 tons of coal per day. The ship would have to be capable of carrying twelve thousand tons of coal. The newly formed Royal Australian Steamship Company commissioned Brunel to build two such ships. Several factors combined to influence the design.

It was known that the larger a vessel, the less onboard space was needed to carry coal and the more room there would be for fare-paying passengers. In 1839 naval architect John Scott Russell had discovered that turbulence and wave-making were a major cause of energy loss. Wave-making could be reduced by giving the ship hollow, sinusoidal lines at the bow and cycloidal lines in the rest of the hull. The length of the bow section (called the "entrance") related to the wave of the required speed. The higher the speed the longer this section and therefore the longer the entire hull had to be. This calculation gave rise to what was known as Russell's "wave-line" principle. So *Great Eastern* was going to be long. At 692 feet length and 82 feet beam, and displacing 32,160 tons, she turned out to be six times larger than any ship built so far. The clear danger of such a design was that in heavy seas when balanced on a single wave amidships or supported by only one wave at bow and one astern a ship of this length might break her back. Brunel dealt with this eventuality by using Stephenson's tubular Britannia Bridge construction. He sandwiched the ship's longitudinal tubes between the inner and outer hulls and linked them with transverse tubes so that the ship resembled a giant box girder. The ship's three million rivet holes were punched out by Roberts's machine.

On January 31, 1858, the sixth attempt to launch *Great Eastern* succeeded. The costs of launch, estimated at £14,000, had risen to £100,000. Other costs had also spiraled and there was talk of auctioning the ship off before her maiden voyage. So far the ship had cost the astronomical sum of £640,000, the owners were already £90,000 in debt, and the ship was only partially engined and still to be fitted out. Then the Indian Mutiny of 1858 disrupted Eastern trade, so an alternative, Atlantic run began to look more attractive. Costs continued to spiral as the elaborate fitting-out took place. Shares in the vessel were now a fifth of face value, debt was mounting and on her maiden voyage *Great Eastern* sailed with forty-six passengers instead of the possible three hundred. She returned with only seventy-two. By 1863 the company directors decided to cease operations. A year later, after failing to meet the reserve price at auction, the ship was sold. In the words of one of the buyers: "Mr Barber went down to Liverpool to attend the sale when, strange to state, a ship that had cost a million of money and was worth £100,000 for the materials in her, was sold to us for £25,000."

Great Eastern now began a new career, thanks to a thirty-four-year-old American self-made millionaire named Cyrus Field,[31] who had **31** 8 *30*
amassed a fortune in paper-making. In 1858 he had succeeded in linking Newfoundland and England with a transatlantic telegraph cable, when on the very day the cable was opened for the first mes-

Fig. 6: *The* Great Eastern, *laying cable over the stern. The ship would end up as a floating music hall and bar.*

sage traffic, it mysteriously failed. Examination showed that manufacture of the earlier cable had been incorrectly done, allowing the copper cable core to protrude through the insulation and come into contact with seawater. New and more accurate measures were suggested for the amount of insulation needed. This left the matter of finding a ship capable of carrying the new two-thousand-mile cable as well as all the necessary equipment for laying it, as well as 120 sheep, 10 oxen, 20 pigs and scores of chickens. Only one ship was big enough: *Great Eastern*. On her first attempt, about one thousand miles out the cable snapped. The final attempt started in July 1866. The ship had been refitted with improved engines and she set off with 2,400 miles of new-design cable weighing only 5,000 tons, as well as 8,500 tons of coal fuel, 500 tons of equipment and assorted farmyard supplies as before. This time the laying was successful. On
32 9 *31* July 26 the first Morse[32] Code message was transmitted.

Key to the entire effort was a new kind of cable insulation discovered in Singapore by a surgeon named William Montgomerie. The material was called "gutta percha," which was produced by evaporating the milky latex of the gutta percha tree into a firm, inelastic material that could be softened and molded in hot water. In cold, high-pressure environments the material became hard but not brittle, so it was ideal for the deep ocean. Gutta was also later used to make rowing boats for Arctic explorers, and in ear trumpets, stethoscopes, domestic telegraphs and speaking-tubes, artificial teeth and fillings, chemical apparatus and machinery drive belts. It was employed as decorative and fine-art material in inkstands, pen-trays, baskets and vases. It also became a substitute for leather, papier-mâché, cardboard, wood, millboard, paper and metal.

Gutta percha was also found to be the ideal material for golf balls. The first golf club was formed in 1744, in Leith, Scotland. In 1754 another club opened at St. Andrew's, and it was here that the eighteen-hole course would be developed. Before this, courses ranged from five to twenty-five holes. At St. Andrew's the course ran alongside the shore: eleven holes out and the same number back. In due course these were reduced to eighteen and this became standard. By the early seventeenth century leather balls filled with feathers had replaced those made of boxwood. However the "featheries" easily became waterlogged and were rarely perfectly spherical. So when gutta percha balls became available they caused a sensation. At first they flew badly, but then it was noticed that after the ball had been hit a few times flight improved. Balls were then beaten with a small hammer to produce the effect deliberately. Today we call this "dimpling." In 1850 the new gutta balls were so much cheaper and long-lasting than their predecessors they created a boom in Scottish golfing.

This was a time of general expansion in Scottish leisure-time activities, following the country's rapid industrial development and the spread of the railroads. Economic growth had accelerated late in the previous century, when manufacturers began to take advantage of Scotland's ideal position for the rapidly growing transatlantic trade. Scotland became the center for tobacco exports, sugar and cotton. Trade generated an infrastructure of banks, warehouses and ports. These in turn facilitated further industrial development, particularly after the 1801 discovery by David Mushet (an assayer at the Calder Iron Works) of large deposits of blackband ironstone throughout the western counties of Scotland. These deposits were rich in iron mixed with coal but required an unprofitable amount of heat to use for smelting.

In 1816 a new gasworks was built in Glasgow. The manager of the works was James Beaumont Neilson, who had previously worked as an engine-wright at John Roebuck's Boroughstounness colliery. Neilson would radically change the face of Scottish industry by making it easy and profitable to use blackband ironstone. In 1820 he became interested in finding a way to sell gas to iron foundries. For years it had been thought that because foundries produced more iron in winter, refrigeration of the foundry air blast should increase output. This proved not to be the case. Experiments convinced Neilson that production would be increased with a hot-air blast. This could be done by first passing the air through a tube surrounding a gas burner, thus adding heat to the fire and saving fuel. Neilson entered into partnership with a local industrialist, Charles Macintosh, and the apparatus was built. The air passed through tubes placed over a gas grate and was heated to 600 degrees Fahrenheit. The hot blast tripled production. More important, the heat was now sufficient to make use of the plentiful blackband ironstone, and since the rock contained both iron and coal, extra fuel was not required for smelting.

In 1830, 40,000 tons of Scottish pig-iron were produced. Only ten years later the use of ironstone increased annual output to a quarter of a million tons. By 1848 Lanarkshire alone had fifteen ironworks and ninety-two furnaces and Scotland was producing half a million tons of iron a year. Cheap iron also launched the Scottish shipbuilding industry. In 1835 5 percent of British shipping was built on the Clyde. Between 1851 and 1870 this rose to 70 percent.

33 114 *214* Charles Macintosh,[33] Neilson's partner in the hot-blast venture, had also signed another deal with the Glasgow gasworks. Two of the by-products of the gas-making process (which involved baking coal to collect the gas it gave off) were coal tar and ammoniac liquid, both of which were dumped in rivers and quarries. Since 1777 Charles Macintosh's father, George, had been manufacturing a dye known as "cudbear." Cudbear gave a violet/purple color to wool and silk, but could with the use of acids also produce red. Alkali would return the color to purple or blue. Cudbear was also used by paper stainers, who knew it as "litmus." The chief ingredients for the dye were lichen and ammonia. Previously the ammonia had been provided from the urine of friends and workers. From 1819 the ammonia was obtained from coal tar. Further treatment of the coal tar also produced naphtha, used by Charles Macintosh to liquefy rubber, which he spread between sheets of cotton to produce the first raincoat (which still bears his name in Britain today).

Earlier George Macintosh had formed a partnership with David Dale to set up a works producing another dye, Turkey Red, but the venture failed and the factory was sold off in 1805. Dale (one of Scotland's foremost textile manufacturers and the founder of the country's first cotton mill) then set up a mill on the River Clyde at New Lanark, and by 1799 it was the largest in Scotland, employing over thirteen hundred workers. That year Dale sold the mill to a Manchester company, which installed a manager named Robert Owen, who then married Dale's daughter. Owen's enlightened social attitude brought new and liberal conditions of work for the employees in New Lanark, where he provided education and medical services for workers and their children. In 1824, when Owen had become a leading light in what would eventually be the socialist movement, he moved to America and set up the utopian commune of New Harmony, Indiana. By 1827 the venture had failed and Owen returned to the United Kingdom.

His four sons remained behind in the United States and became
citizens. The eldest, Robert Dale Owen,[34] went on to follow in his fa- **34** 20 *44*
ther's liberal footsteps, first in the Indiana legislature and then in **34** 130 *248*
Washington D.C., championing causes such as contraception and the emancipation of women. A pamphlet he wrote on birth control, titled "Moral Physiology," was published in 1830 and provided much (unacknowledged) material for "The Fruits of Philosophy" by
Charles Knowlton[35] which, when it was republished by Annie Besant **35** 20 *44*
in England, led to her trial for obscenity.

In 1846 Representative Owen introduced a bill to the U.S. Congress to authorize acceptance of a large foreign bequest to America. In the curious way history works, the bequest linked Owen through

his father-in-law, David Dale, to the Glasgow gasworks manager
James Neilson. Neilson's employer at one time had been the entre-
36 17 *40* preneur John Roebuck,[36] whose interest in coal mines had led him to
37 16 *39* become acquainted with James Watt,[37] who was working on ways to
37 139 *259* make the Newcomen steam pump (used primarily for draining
mines) more efficient. Roebuck offered Watt a cottage in the grounds
of his house near Edinburgh so that Watt could experiment further.
He also took over Watt's debts in exchange for a two-thirds share of
any profits that might come out of the experiments. In 1772 Roebuck
38 18 *40* sold his share to Matthew Boulton,[38] a metal manufacturer of Bir-
38 97 *178* mingham who would in 1774 become Watt's engine-maker.

Nearly ten years later, and by now rich enough to indulge in scientific work of a more theoretical nature, Watt wrote a paper about his experiments on the composition of water. This brought him into conflict with Lord Henry Cavendish, who claimed to have made the same discovery (that water was made up of two parts hydrogen and one part oxygen) before Watt. The argument for priority rested on a number of mistakes made by the Royal Society regarding the date of publication of various documents and the dates on which various letters had been received. In the long run in 1785 when Watt became a fellow of the Royal Society he met Cavendish and the two men resolved the issue amicably, realizing that they had both been working on the same problem independently.

The previous year Cavendish had been on a geology field trip to Fingal's Cave in Scotland together with a young assistant named James Macie. Macie was the illegitimate son of the duke of Northumberland and an avid amateur scientist. On Macie's graduation from Oxford in 1786 Cavendish proposed him for membership in the Royal Society, and Macie began a career whose high points were the discovery of a flintlike substance in the joints of bamboo, the discovery of a new way of making coffee (that foreshadowed the modern vacuum method), the study of tears, and the identification of a type of calamine.

Macie's illegitimacy severely curtailed his activities, as witnessed by the British document of naturalization granted when he was born in France: He "shall not be hereby enabled to be of the Privy Council or a member of either House of Parliament or to take any office or place of trust either civil or military or to have any grant of lands, tenements or hereditaments any inheritable property from the Crown to him or to any person or persons in trust for him anything herein contained to the contrary notwithstanding." The document may have been instrumental in establishing one of the greatest scientific institutions in the world, because in the event that no blood relative surviving him should have heirs Macie's will left his entire fortune (equal to 104,960 gold sovereigns) to the United States of America. This bequest was the subject of Robert Dale Owen's bill of 1846.

The gold had arrived earlier in 1838 when it had been recoined as over half a million gold American dollars (approximately two billion dollars in modern money). Then, in one of the more complex and devious financial deals in the history of America nearly $550,000 of the funds were almost immediately invested in the Real Estate Bank of Arkansas (which delivered a low rate of return) to be redeemed only in 1860. Subsequent investigations in 1845 revealed that the bank had grossly overvalued its real estate, was in serious danger of foundering and had in the meantime not paid any of the interest due on the invested sum. After heated discussions on Capitol Hill it was agreed that the Treasury would guarantee the missing bank interest. The way was clear for Owen's bill of acceptance.

Only now could the bequest be used for the purpose Macie had originally intended in his will of 1826: the foundation in Washington of "an establishment for the increase and diffusion of knowledge among men." The will also stipulated that the institution be named after Macie himself. Since on his father's death Macie had been given permission to take on the family name of the dukes of Northumberland (Smithson) the new American institution was duly named The Smithsonian.

CHAPTER 3

Drop the Apple

The Smithsonian Institution has well exceeded James Smithson's hopes. It has been at the forefront of scientific research since its foundation over 150 years ago. Today it includes sixteen museums and galleries and the National Zoo. It operates facilities in eight U.S. states and in Panama and has research teams in the field all over the world. One of the Institution's galleries contains an unparallelled collection of crystals, including the one investigated by the Institution's founder in 1801. Smithson described and classified the three separate substances known as calamine: carbonate of zinc, hydrous silicate of zinc and zinc oxide. In honor of his work, in 1832 the carbonate form was given the name "smithsonite." Then in 1852 the name was applied instead to the silicate. Today it is also used for the zinc oxide. But whatever its composition, calamine is now best known for use in lotion, sunburn cream, cosmetics, mineral supplements and in some cases as a treatment for chickenpox. It is also used in ceramics, as a rubber reinforcer, in semiconductors and as a photoconductor in photocopying machines.

What is less commonly known about calamine is something that was first noticed in 1703 by an unknown Dutch jeweler who reported that after heating the crystal tourmaline on coal embers, as it cooled small particles stuck to it as they would to a magnet. Tourmaline became known as the "electric stone." In the early nineteenth century the French researcher René-Just Haüy discovered that the crystal behaved like this because it carried an electric charge strongest at each pole and that if tourmaline were smashed, each fragment would exhibit the same bipolar characteristics. Haüy also discovered the same "pyro-electric" features in other crystals, including calamine. Then it was discovered that the phenomenon was associated with a change in the crystal shape during heating and cooling.

In 1880, two French scientists, Pierre and Jacques Curie, found that pressure on the crystal would produce the electric charge, and that the charge was proportional to the amount of pressure. They named this phenomenon "piezo-electricity" (Greek: *piezen*, to press). A year later, using both quartz and tourmaline, they proved the reverse: an electric charge changed the crystal shape. The Curies went on to design an instrument for measuring very small electrical charges by the amount of shape-change the charges caused in a crystal. The crystal they used was quartz, and the instrument became known as a piezoelectric quartz balance. The device was the subject of Jacques's doctoral dissertation in 1889.

In 1894 Pierre Curie met a twenty-seven-year-old Polish physicist who would use the balance to extraordinary effect. Her name was Marie Sklodowska, and a year later she and Pierre married and began the work that was to change the world. By late 1897 Pierre was teaching an electricity course at the Paris Municipal College of Physics and Chemistry. Two years earlier Wilhelm Roentgen had discovered X-rays, and a few weeks after that French research Henri Becquerel attended a meeting at the French Academy of Sciences and heard that the new rays caused phosphorescence on the glass wall of a vacuum tube. Since Becquerel's father had done research on phosphorescence, Becquerel began to investigate the possibility that the X-rays Roentgen had discovered might be caused by phosphorescent materials.

At one point in his experiments Becquerel used powdery white uranium salts: potassium uranyl disulfate. After wrapping a photographic plate in thick black paper to keep out the light, he laid a saucer on which some of the salts had been placed on top of the paper and then left everything exposed to sunlight for several hours. When he developed the photographic plate the outline of the salts was clearly visible, together with the outline of any object placed between the salts and the paper wrapping. Assuming the sunlight was causing the salts to fluoresce and generate the image, Becquerel prepared another experiment. For a number of cloudy days without sun the package (the wrapped plate, a copper cross and the salts) remained in a closed cupboard awaiting better weather. At one point Becquerel decided not to wait any longer and developed the photographic plate. To his surprise the cross was clearly visible in the picture even though there had been no exposure to sunlight. In May 1896 Becquerel announced the news and dropped the matter.

Nine months later Pierre and Marie Curie's quartz balance identified a small electrical charge in the air above the same uranium salts. Intrigued, Marie set out to discover if any other substances behaved the same way. On February 17, 1898, she tested a sample of pitchblende and discovered the charge it produced was much greater than that from the uranium salts. The Curies began boiling down and distilling the pitchblende to find out what was causing the charge. By late June their highly concentrated sample was giving off a very large charge. The Curies described the substance as "400 times as *active*" as uranium, and in July, coined the term "radio-active." It was this radioactivity that had been causing the image on the photographic plate. In December 1898 they named their new substance "radium."

One of the Curies' close friends (so close that after Pierre's death he would briefly become Marie's lover) was Paul Langevin, a brilliant physicist. Pierre had been Langevin's lab supervisor at the Municipal College and Langevin had been present at the start of the Curies' radioactivity work. In 1914, at the start of World War I, Langevin was working on ballistics when he was asked to look into the problem of submarine detection. Langevin turned to the old Curie quartz balance. In less than three years he produced what his lab team referred to as the "Langevin sandwich": two three-centimeter layers of steel sandwiching a four-millimeter layer of quartz crystal.

When an electric charge of the right frequency was applied to the quartz it caused the crystal to change shape and eventually to oscillate at its own resonant frequency. This in turn caused the steel plate to vibrate. When the "sandwich" was set into the hull of a ship the vibrating steel would send out high-frequency pulses into the surrounding water. If these vibrations hit an object, they would bounce back and be picked up by a receiver, which would apply the returning vibrations to another quartz plate. The effect of these vibrations would cause the quartz to emit electrical signals that could be processed to show the range and size of the object. In 1918 the system identified a submarine from 600 meters away, even when the submarine was stationary on the seabed. British and American researchers took the technology (now known as sonar) to the same stage, then the war came to an end before ship-fitting could be begun, and in the 1920s all three powers dropped the matter.

In World War II German U-boat activity brought sonar to the fore again. By the end of May 1940 the wolfpacks had sunk 241 ships totalling 853,000 tons. In June the total was 58 ships, in July 38, August 56, September 59 and October 63. In the same period the Germans had lost only six of their 57 U-boats. Sonar was rapidly deployed, and it helped the Allies win the Battle of the Atlantic. By the end of World War II U-boats had sunk a total of 23,351 Allied ships, and 782 U-boats had been destroyed.

In the early years of the war, before America joined, U-boats were sinking ships faster than British yards could build them. In February 1941, President Roosevelt announced an emergency shipbuilding program. He described the new ships to be built as "dreadful looking objects." Each was to be just over seven thousand tons, and the aim was to produce them quickly and cheaply. Many of the ships were launched with no radio direction finders, fire detection equipment, emergency generators or lifeboat radios. Some went to sea with only one anchor. A total of 2,710 of these "instant" ships were built.

The first one, the *Patrick Henry,* was launched on September 27, 1941, to the president's ringing words: "Each new ship strikes a blow at the menace to the nation and for the liberty of the free peoples of the world." The ships became known as "Liberty Ships." Their rate of manufacture was phenomenal. By September 1942, American shipyards were launching three ships a day. The *Robert E. Peary* was launched on November 12, 1942, only four days, fifteen and one-half hours after her keel had been laid. Liberty yards assembled each ship from thirty thousand components that had been mass-produced at thousands of factories in more than thirty-two states. In those places without slipways the yards simply flooded the dock and the ship floated out. Component parts were consumed virtually nonstop and it was quite commonplace to see complete deckhouses erected upside-down on a wheeled trolley and then inverted and placed in position. Stockpiles of double-bottom hull sections, with piping already installed, waited to be dropped complete onto the keels.

One reason for such a high rate of production was the early decision to make Liberties virtually all-welded. Oxyacetylene welding was a simple "fusion-welding" process. Both pieces of metal were heated to 3,100 degrees Centigrade, at which point they melted and flowed together once filler material had been placed in the gap be-
39 57 *119* tween them. Acetylene[39] gas had originally been discovered in 1836
39 127 *244* by Edmund Davy, cousin of Sir Humphrey. The technique for pro-
ducing it on a commercial scale was discovered by accident. In December 1892 a French researcher named Henri Moissan (the man who provided Marie Curie with uranium salts during her early experiments) was attempting to make artificial diamonds in an electric-arc furnace in which two blocks of lime held a crucible, heated when the spark between two carbon electrodes caused them to burn incandescent and produce extremely high temperatures. Moissan failed to make diamonds but went on to other experiments, out of which came a material second in hardness only to diamonds: calcium carbide. Pouring water on calcium carbide produced considerable quantities of acetylene gas.

Moissan had developed his electric-arc furnace from lighting technology. In 1809 Humphry Davy had made the first arclight by attaching a charcoal rod to each terminal of a Volta battery, bringing the ends of the rods close together and producing a brilliant white light as the current jumped the gap between them and the rod tips burned away. From 1845 onward determined efforts were made to establish arclight on a commercial basis. However, the arclight required a constant supply of electricity, and that required the development of a generator.

In one of those related-yet-unrelated incidents with which history is replete, other people were trying to find ways of generating large quantities of hydrogen to burn in the limelight apparatus used by lighthouses. One way of doing this was to disassociate hydrogen molecules from water with an electric charge. This led to the requirement for a constant supply of electricity, which in 1870 led to the development of the Gramme dynamo. The dynamo was exactly what arclight makers needed, and the arclight soon became commonplace in railroad stations and lighthouses.

The other obstacle to arclight production was the need for a regulator to advance the carbon rods toward each other at a precise rate so that they were always the optimum distance apart. If the rods came too close, they burned out too quickly; if they were too far apart the light level was insufficient. An English electrical engineer, W. E. Staite, developed a clockwork regulator. His carbon rods were placed vertically, one above the other. A wound-up, weight-loaded gear train was released (by the expansion of a copper wire heated by the arc), and it engaged with rackwork to raise the lower carbon rod gradually as both rods burned away. In 1849 a French scientist, Leon
Foucault,[40] developed an improved regulator, which worked well **40** 119 *221*
enough for arclights to make their historic entry to the theater (where they replaced limelight).

Foucault's interest in the regulation of motion stimulated his most spectacular invention. In 1851 he hit the headlines when in the Paris Pantheon he hung a two-hundred-foot steel piano wire pendulum to which he attached a sixty-two-pound cannonball with a stylus fixed to its underside. He pulled the ball to one side and tied it off with a thick thread. When he set fire to the thread it parted and the pendulum was released unaffected by any influence. As the ball swung back and forth, the stylus traced out a line in sand on the floor. At the end of an hour the stylus had traced a line that had gradually turned through 11 degrees 18 minutes. Foucault's pendulum was the first demonstration of inertial motion, in which the pendulum swung unaffected by the rotation of the Earth beneath it (as shown by the stylus). This was the first physical proof that Copernicus had been right.

Following his inertial work, in 1867 Foucault produced a heliostat and siderostat, two mechanisms designed to operate by clockwork so as to keep a telescope pointed at the sun or a star. Earlier, in 1845, Foucault had become interested in this aspect of astronomy following the announcement of a new photographic technique by Daguerre, and had taken the first daguerreotype picture of the sun, confirming the existence of limb-darkening as well as revealing several sunspots and their penumbrae. A daguerreotype taken during an eclipse in 1850 also revealed solar prominences and the sun's corona.

By 1819 Louis-Jacques-Mandé Daguerre was a well-known stage designer at the Paris Royal Academy of Music, and in 1822 he began to present spectacular dioramas, complex visual shows that included effects produced by the sequential illumination of scenes painted on multiple semitransparent backdrops, so that the scenes "dissolved" from one to the other. Dioramas were the rage of Paris and London, and it was probably during the drawing and painting involved in preparing the backdrops that Daguerre became acquainted with the
41 99 *182* painter's standard tool: the camera obscura.[41] In a camera obscura, light entering a pinhole in a darkened box would show an image of the outside world, upside down, on the opposite side of the box. The technique had been in use since the seventeenth century when it had aided accuracy of reproduction by artists such as Dürer and by as-
42 100 *182* tronomers such as Kepler,[42] who used one to make drawings during
42 108 *207* a partial solar eclipse in 1600. Daguerre decided to use the camera obscura to develop a photographic process.

In 1831 he learned from the earlier work of Joseph Niepce that iodide of silver would very slightly darken in the presence of light. He exposed a silvered copper plate to a camera obscura image and then placed the plate in iodine vapor. The image it created was too faint to see. One day in 1835 Daguerre put away an exposed plate in a cupboard full of various chemicals, intending at some point to polish the plate and use it again. A few days later he opened the cupboard and to his amazement found the plate now carried a distinct picture. By process of elimination he discovered that the image had developed

Fig. 7: *An early daguerreotype. The subject is the son of astronomer Sir John Herschel.*

because of the presence in the cupboard of a few drops of spilled mercury.

By 1839 he felt confident enough to make his new photographic technique public. The photographic plate was a thin film of polished silver on a copper base, sensitized by being placed face down in a vessel containing a few particles of iodine whose vapors formed a yellow layer of silver iodide less than a micron thick on the silver surface. After subsequent exposure for a suitable length of time in the camera obscura the plate carrying the latent image was developed by a vapor of mercury heated to 75 degrees centigrade. The mercury vapor stuck to the parts of the plate carrying an image while leaving untouched the unexposed parts. The image was then fixed by immersion in a solution of hyposulphite, which dissolved the unused silver iodide. Finally the plate was rinsed with hot distilled water. The final image was seen immediately as a positive, when examined at such an angle that a dark background appeared reflected in the undeveloped parts of the plate.

The iodine Daguerre used had originally been discovered by another Frenchman as the result of another accident. During the Napoleonic Wars at the beginning of the nineteenth century the Al-
43 10 *34* lies imposed a total trade blockade[43] on all French ports. In conse-
43 59 *120* quence, one of the imports France could no longer obtain from
43 112 *212* abroad was saltpeter. This was unfortunate during a war since saltpeter was an essential ingredient of gunpowder, which was made from 75 percent saltpeter, 15 percent charcoal and 10 percent sulphur. Previously France had imported saltpeter from India and North Africa, where climatic conditions (hot and wet, followed by hot and dry) favored its production. Saltpeter was the product of a process in which organic materials, principal of which were human and animal waste products, decayed in the soil. This decay formed nitrates, which, when combined with potassium compounds, produced potassium nitrate in solution. In the dry season this solution rose to the surface and evaporated, leaving behind a deposit of potassium nitrate salts (saltpeter).

Denied imports by the blockade, Napoleon's scientists set up niter beds all over France and began to produce nitrates artificially by mixing urine and dung with powdered limestone. This made calcium nitrate through the oxidation of the ammonia produced by the breakdown of the organic materials. Turning calcium nitrate into saltpeter involved dissolving the nitrate in water and boiling it in a vat containing potassium carbonate derived from wood ashes. The only problem with this method was that the wood for the wood ash was ruinously expensive. It was also extremely difficult to obtain, since severe government restrictions limited the use of forests to shipbuilding.

In 1811 Bernard Courtois, a French chemist who operated niter beds near Paris, hit on the idea of producing ash by burning a type of seaweed known as "kelp." Before it could be used the kelp underwent an intermediate process to extract sodium carbonate for use in the soap industry. This took place in copper vats in which the sodium carbonate extraction process left a thick insoluble deposit. On one occasion when Courtois was cleaning the deposit out with sulphuric acid he used too much acid and saw violet vapors rising from the vat. Then he found crystallized violet deposits on the inside. He passed these crystals to a couple of chemist friends and in due course in 1814 the crystals were identified as a new element and given a name based on the Greek word for violet (*iode*): iodine. Unfortunately for Courtois, in 1815 the war and the blockade ended, France returned to importing cheap foreign gunpowder, and his business went bust.

The kelp Courtois used had been harvested along the coastline of northwestern France. A much larger kelping industry had already developed in Britain, in response to earlier industrialization. Most of the British kelping industry was concentrated on the western coast of Scotland, where villagers raked the seaweed off rocks and burned it in circular pits about five feet across and one foot deep. A fire was then lit with dry kelp in the center of the pit and wet kelp heaped on top. The dense white smoke given off by pits could sometimes be seen miles away at sea and kelp-producing islands often gave the appearance of volcanoes. As the kelp burned it solidified and was beaten flat with shovels, and flat stones were placed on top to press it down. After two days the burned kelp had fused into a single large piece of slag, which was then broken up into smaller stones and ground down. So great was the demand for kelp ash that Scottish landowners made fortunes. Since the Union with England in 1707 the Scottish aristocracy wanted money to maintain a fashionable presence in London or Brighton. The second Lord Macdonald of the Isles had an annual kelp income of twenty thousand pounds. Macdonald of Clanranald (who moved villagers' homes if they spoiled his view) made eighteen thousand pounds. These sums were extremely large at the time, and much of the money went on a luxury lifestyle and the construction of pseudo-Gothic castles, which can still be seen today.

The demand for kelp was widespread. Although it was not known at the time, kelp contains as much nitrogen as animal manure, and it was prized as a local fertilizer. Industrial uses for kelp were more profitable and demand grew steadily through the mid- and late-eighteenth century. The earliest commercial demand for kelp ash had come from glassmakers, denied (by the same shipbuilding constraints) the wood ash they had traditionally used to make "frit," an essential ingredient in their product. Kelp made the greatest contribution to British economic advance in the textile industry. When kelp ash was dissolved in water to which quicklime was added, the alkaline solution could be used as the first stage in bleaching cloth. It could also be mixed with fatty substances to make soap (to be used in washing grease out of raw wool before it was woven). Muriate of potash could also be extracted from the mixture and used in the dyeing industry. The greatest demand for kelp ash was, however, generated by the extraordinary rise in demand for cotton.

Middle Eastern cotton was brought to England in 1601, probably by Dutch immigrants. At first it was used to make fustian, a linen-cotton mix. Around 1730 cotton stockings became popular. At the same time the printed calicoes (cotton products named after Calcutta, the Indian port of export) and chintzes being brought into Eu-
44 66 *137* rope by the Dutch East India[44] Company were creating an insatiable
44 109 *208* market. These imports were met with some opposition by local English manufacturers who described the new cloth as "a tawdry, pie-spotted, flabby, low-priced thing called Callico . . . made by a parcel of Heathens and Pagans that worship the devil and work for a half-penny a day." Cotton was durable and comfortable to wear, it washed and ironed well, and it could be dyed and worked with ease. As the standard of living rose cotton became popular for aristocrats' underwear and workers' shirts. The exceptionally good weather and bumper harvests in the first decades of the eighteenth century had brought a fall in the price of food and a rise in the value of wages. In consequence the birth rate rose and the market for household goods and clothing expanded rapidly. Since England ruled India at the time regular cotton supplies seemed assured. The requirement only remained for technology that would enable cotton manufacture to keep up with demand.

In 1753 a woolen manufacturer named John Kay invented a "flying shuttle" that could be tugged back and forward across the loom on small wheels. It doubled the weaver's output. The new loom worked faster than thread-spinners could keep up so in 1764 weaver James Hargreaves developed the "spinning jenny" with which one spinner could spin thread onto a number of spindles at the same time. In 1779 an inventive wigmaker named Richard Arkwright automated the thread-spinning process. Sets of rollers moving at different speeds pulled the thread from the mass of raw cotton, then automatically twisted it and wound it onto spindles. Arkwright drove his machine by water power and took cloth production forever out of the cottage and into the factory. In his first textile mill the machinery, filling six floors, was run by belt drives set on shafts turned via gearing by one huge waterwheel. Arkwright called his device a "water frame." This met all the demand for coarse cloth, but for finer fabrics the manufacturer had still to use the older (and now slower) jenny.

In 1779 Samuel Crompton combined the principles of the jenny and the water frame in one hybrid machine, which he aptly named the "mule." The thread was wound onto bobbins in the same way as on the water frame, but the spindle-mounting moved away and back, stretching the thread as it was twisted. This resulted in a much thinner thread, making possible the production of fine muslin cloth that was washable and now cheap. By 1785 the *Annals of Commerce* wrote: "Women of all ranks from the highest to the lowest, are clothed in British manufactures of cotton, from the muslin cap on the crown of the head, to the cotton stocking under the sole of the foot . . . with the gentlemen, cotton stuffs for waistcoats have almost superseded woolen clothes . . . cotton stockings have also become very general for summer wear."

In 1787 there were 119 cotton mills in Britain. By 1837, 1,791 mills employed a quarter of a million people. In the same period, cotton output rose from one to ten million pounds a year. Then in the latter part of the eighteenth century war in India began to interrupt cotton supplies. Indian sources had always been unreliable. Thanks to the dreadful state of the roads, loads often arrived at ports too late for shipment. As the British consumer became more sophisticated, the short-staple coarse Indian fibers lost their appeal. Shortage of Indian producer capital also drove down quality, and with most growers incurring debts to be paid off by their children and grandchildren there was no money to spare for better husbandry. By 1792 the situation had reached crisis point. Some more reliable source of cheap, good-quality cotton had to be found.

That year a young Yale graduate named Eli Whitney, on board ship from New York to Georgia, fell in with a fellow passenger who happened to be the wealthy widow of a Revolutionary War general, Nathaniel Greene. She invited him to visit her on his way to Savannah. At her Mulberry Plantation he met many of Widow Greene's friends and heard constant talk about the problems associated with removing seeds from the green-seed cotton grown all over the South. It took one man one day to clean the seeds from one pound. Whitney said he thought he might be able to design something to deal with this problem. Mrs. Greene promptly offered him room and board, and he stayed on.

The following year Whitney applied for a patent for his new cotton gin. It was a simple apparatus consisting of a wooden cylinder with wire teeth set in transverse rows around it. When the cylinder was turned the wire teeth picked up raw cotton from a hopper and pulled it through a wire screen that separated lint from seed. At this point a revolving brush removed the seeded lint from the wire teeth. Whitney's gin was able to seed cotton a hundred times faster than a man, and it would one day be a major contributing factor in the outbreak of the Civil War, because the gin made cotton profitable enough to sustain the social system that believed in slavery. By 1807 U.S. cotton exports had risen from 190,000 pounds in 1791 to over 66 million pounds. By 1825, with the British market avid for every pound of cotton it could buy, thousands of slaves were cultivating cotton on hundreds of thousands of acres throughout the South. In Britain 450,000 factory workers were processing and weaving the raw import. By 1859 two-thirds of the world's cotton was American, and Senator Hammond of South Carolina was uttering the immortal words: "Cotton is King!"

Unfortunately, all this did little for Whitney's fortunes. Pirated copies of his gin appeared everywhere and redress was virtually impossible. As he said: "I have a set of the most depraved villains to combat and I might almost as well go to Hell in search of Happiness as to apply to a Georgia Court of Justice." Whitney got out of the ginning business and took up making muskets with interchangeable parts. In January 1798, back in New Haven, Connecticut, he won a contract with the U.S. government to deliver four thousand guns within twenty months and another six thousand the following year. In the end it would take him nine years to fulfil the contract. In the course of this Whitney invented the milling machine, which enabled workmen to cut metal to standard patterns, so as to make the interchangeable parts from which every musket was to be made. Another Connecticut Yankee, Eli Terry, was doing the same thing with clocks. There are suggestions that the two men met, but it is more probable that word of Whitney's techniques had gotten around. After all, he had been copied before.

Terry started by making his wooden clock parts with a hand-operated wheel and pinion-cutting engine, but by 1806 he had signed a contract for six thousand clock movements and needed to move on to mass production. By 1820 thirty workmen were using templates to make twenty-five thousand identical wooden clocks a year and Terry was rich. In 1816 he was joined by a young cabinet-maker named Chauncey Jerome, who had been hired to make Terry's clockcases. When his contract with Terry was ended Jerome sold Terry his house in Plymouth for a hundred completely fitted mantle-clock movements, made 114 more, encased them all, sold these 214 complete clocks for a house, a barn and seventeen acres in Bristol, Connecticut, and set up in business.

In 1838 Jerome (or his brother) invented the brass clock. The advantage of brass was that it did not need seasoning as did wood, so no delays were involved in going from raw material to finished product. In 1844 Jerome moved to New Haven to make cases, while in his Bristol factory three workmen continued to turn out all the wheels for 500 clocks per day. A single machine using three cutters performed three sequential operations: simple cutting, rounding off teeth and finishing. By 1850 Jerome had two factories in New Haven turning out 280,000 clocks a year. During its operational life the firm exported millions of clocks to Europe, Asia, South America, Australia and the Middle East. However, in 1855 Jerome found himself in financial difficulties. It was at this point he met an extraordinary character named Phineas T. Barnum and was persuaded to go into business with him.

Barnum had begun life as a store clerk, and then set up his own fruit and confectionery shop in Bethel, Connecticut, where he also ran a local lottery. In 1831 he began publication of a weekly newspaper, *The Herald of Freedom,* which he edited for three years until forced to resign after libeling a deacon of the local church. In 1834 Barnum moved to New York and began his show-business career, exhibiting freaks and curiosities all over the country. In 1842 he discovered a two-foot, one-inch midget in Bridgeport, Connecticut, dubbed him "Tom Thumb" and made a fortune showing him off to European royalty and American audiences, who thought the midget a sensation. In 1851 Barnum bought a museum in New York and, in his own words, "scoured America for industrious fleas, automations, jugglers, ventriloquists, living statuary, tableaux, gypsies, albinos, fat boys, giants, dwarfs, rope dancers, dioramas, panoramas, models of Niagara, Dublin, Paris and Jerusalem, Punch and Judy, fancy glass-blowing, knitting machines, dissolving views, and American Indians." Barnum would eventually go into the history books with the "Greatest Show on Earth" circus he started in 1876.

Meanwhile, in 1847 he returned from a trip to Europe, decided it was time to become respectable and built a mansion in Bridgeport. Called "Iranistan," the house was modeled on the Royal Pavilion in Brighton, England: a turreted Oriental palace surrounded by gardens and fountains. By 1849 his New York Museum was staging plays and other cultural events, and Barnum was giving lectures on the value of temperance. In this new mood of sobriety he hit upon another way to make a fortune. He would invite the new Swedish soprano Jenny Lind to tour America.

At the time Lind was twenty-nine and already the toast of Europe. Although no beauty, she had tremendous personality and the voice of an angel. As Barnum said to a reporter: "It is a mistake to say that the fame of Jenny Lind rests solely upon her ability to sing. She is a woman who would have been adored if she had had the voice of a crow."

After a few years' training in Paris Lind had returned to Stockholm in 1842 to sing Norma. People cried at the sound of her voice. Hans Christian Andersen fell madly in love with her but his feelings were not reciprocated. In 1844 Lind sang Norma in Berlin. It was her first role outside Sweden. She was an immediate sensation and over the next six years became famous everywhere. Her concerts were booked out weeks in advance. Mendelssohn idolized her. She was waited on hand and foot by princes and empresses. In 1847, at her London debut, Queen Victoria dropped her bouquet at Lind's feet. Her voice was superb, able to manage crescendos better than any other singer alive and capable of expressing magnificent emotional intensity.

When Lind received Barnum's offer to tour the States she was wary. His reputation for showmanship and extravagant public relations tricks was already well established. However, Lind accepted because

Fig. 8: *A typical piece of P. T. Barnum pizzazz: the daring spectacle of an all-woman band in the streets of New York.*

Barnum had converted everything he possessed to cash, taken out mortgages on his property and obtained loans from friends so as to be able to give Lind an advance of $187,500 (today, the equivalent of over $2 million).

Lind arrived at the dockside in America to a Barnum-orchestrated welcome. Flags and triumphal arches decked the port, and thanks to thousands of publicity handouts twenty thousand fans waited outside her hotel. After a rapturous opening night in New York Lind toured the American East Coast and then moved on to California. The general public went crazy. One man offered to pay a thousand dollars to touch her shoulder "and see where the wings began." Another paid $650 for a ticket even though he could not attend the concert. The American tour made Lind rich enough to buy herself out of the contract with Barnum after ten months and return to Europe.

Earlier in 1847, Lind's first triumphant performance at Her Majesty's Theater in London had brought her into contact with another musical megastar, Giuseppe Verdi, who had written the opera for that occasion. The work was *I Masnadieri,* based on a work by Schiller. Lind sang the leading role of Amalia and Verdi himself conducted the first two (of only four) performances. In the event Verdi and his star did not get on. Verdi's amanuensis Muzio (who accompanied him to London where they both complained about the weather, the food, and the English audiences) wrote : (Lind) "is inclined to err in using excessive *fioriture,* turns and trills, things which were liked in the last century but not in 1847." Verdi had accepted the commission to write *I Masnadieri* because he could earn four or five times as much writing for foreign opera houses as he could writing for La Scala in Milan.

The theme of the London opera was for Verdi a typical story of the fight against cruel authority and the triumph of heroism against all odds. Living in Italy under Austrian occupation, Verdi risked life and limb with operas whose themes tended toward thinly disguised nationalist propaganda. *Nabucco* was about Hebrews captive under the Egyptian yoke, and at its first performance had set off a near-riot. *The Battle of Legnano* featured Barbarossa defeating the Lombard League (opening chorus: "Long live Italy!"). *A Masked Ball* had been intended as the story of the assassination of King Gustav III of Sweden, but under pressure from the censor the story was eventually set in pre-Independence Boston.

It may have been these nationalistic themes that first brought Verdi to the attention of the khedive of Egypt, Ismail Pasha, whose country was part of the Ottoman Empire at the time and who was looking for ways to loosen the Turkish grip. Ismail invited Verdi to write an opera to be performed at the Cairo Opera House in celebration of an event the khedive had arranged, which was intended to be Egypt's greatest contribution to civilization since the Pyramids. The financial offer for the opera (over seven times the standard rate at La Scala) was too good to refuse, and Verdi dutifully obliged with what turned out to be his greatest and most successful work: *Aida* (sure enough, set to another nationalist theme). In the event *Aida* was not ready in time for the pasha's great event in 1869: the opening of the Suez Canal.

Since Roman times there had been several attempts at building a canal to link the Mediterranean with the Red Sea. In the eighth century the Arabs had given up because of the risk of opening their seas to the Byzantine navy. In the fourteenth century the Venetians decided the enterprise would cost too much. The English were against the idea because the route around Africa was "comprehensive, safe, bold, truly English." Nonetheless a Suez Canal was attractive because it would save the four-thousand-mile journey around Africa. In 1800 Napoleon stimulated new interest in the canal when his engineers surveyed a possible route during his brief occupation of Egypt. With Napoleon's defeat the idea went back on the shelf until renewed French interest (and their rivalry with the British) was expressed by a young diplomat named Ferdinand de Lesseps,[45] French 45 92 *164*
vice-consul in Egypt in 1832, after he read the original report prepared for Napoleon. De Lesseps proposed the idea to the Egyptian ruler Mohammed Ali and received a concession to build. This was due almost entirely to the fact that it had been De Lesseps's father who at the behest of Napoleon had put Mohammed (an illiterate but

charismatic soldier) on the Egyptian throne. The De Lesseps family were royal favorites. The fact that De Lesseps was related to the
46 120 223 French empress Eugénie[46] also helped.

By the time Ferdinand finally began work on the project in 1856 Mohammed's son (and Ferdinand's personal friend) Said Pasha was on the throne. The construction of the canal eventually took twenty-four years, employed twenty-five thousand laborers and used the new mechanical dredgers. The opening in 1869 was attended by dignitaries from all over Europe and America. They included crowned heads, artists and writers, ambassadors and aristocrats. Eight thousand guests sat down to dinner on the night of the opening. As it happened the guests were celebrating something that already belonged to them rather than to Egypt since the expenditure involved in building the canal had bankrupted Said Pasha and obliged him to sell all his shares to the English prime minister, Benjamin Disraeli.

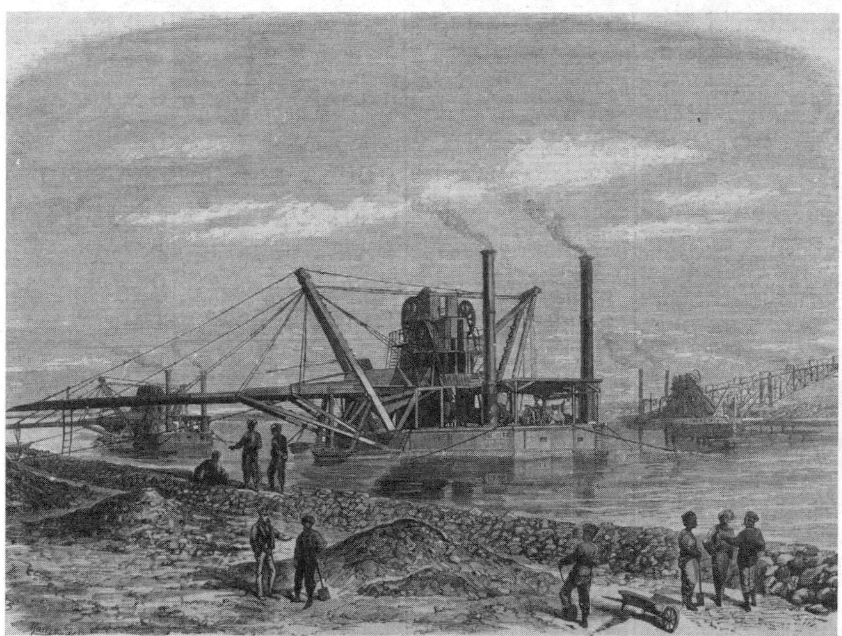

Fig. 9: *Building the Suez Canal. The final breakthrough to the Red Sea was made on August 11, 1869.*

The rest of the shares had already been sold to Switzerland, Italy, Spain, Holland, Denmark and France.

There were one or two among the French guests who might have thought De Lesseps had unfairly taken all the credit for the great achievement. Back in 1833, a year after obtaining the original concession, De Lesseps had met (and helped to keep out of jail) a strange French individual named Prosper Enfantin, who was on a visit to Egypt looking for a wife. Enfantin was a banker and the head of a religious sect known as "New Christians." This quasi-communist, free-love group preached a new kind of social religion and had already set up a number of churches throughout France. Although in the end Enfantin would go to jail for advocating free love in 1833 he and his followers were still active and trying to find him a bride from the East so as to achieve the mystic union between East and West that was part of the sect's attempts to unite all peoples in a brotherhood of love and equality. Enfantin later claimed that part of this union would involve the building of a Suez Canal.

Enfantin's interest in canals had been stimulated by the man who also inspired him to become leader of the New Christians: Henri de Saint Simon. At the age of nineteen Saint Simon had fought on the American side during the War of Independence. Of his presence at the siege of Yorktown he said later: "I contributed in a rather important manner to the capture of General Cornwallis and his army. So I may regard myself as one of the founders of the liberty of the United States." Saint Simon also claimed that in 1783 while in Mexico he had proposed a precursor to the Panama Canal. In 1787 he went to Spain to outline his plan for a canal from Madrid to the Mediterranean. Later he developed plans to unite the Danube and Rhine rivers and to link the Rhine with the Baltic Sea. After a period in 1795 when he became one of the leading financial figures in Paris, by 1797 Saint Simon had failed in several business ventures and was practically destitute. It was at this time he decided to become a philosopher.

Living in increasingly dire circumstances, often penniless and relying on the charity of friends, Saint Simon began to conceive of a grand new social order. In 1817 he started publication of *L'Industrie,* a periodical in which he expounded his new theory: that society depended entirely on industry; that it was industrial "producers" who maintained society; and that political power should rest in the hands of financiers and industrialists. By 1821 he included spiritual values: all men were brothers, and spiritual power must spring from a scientific understanding of the world. Not surprisingly, the new view gained adherents among businessmen, bankers and engineers. The "positive" scientific view of the value of human knowledge in ameliorating the condition of society and the importance of applying scientific analysis to how society functioned has since earned Saint Simon the title of "father of sociology." In 1823, aged sixty-two and once again destitute and depressed, he fired seven bullets at his head. He lost an eye from one of the bullets. The rest missed and Saint Simon survived. Two years later he had founded New Christianity, and shortly thereafter was dead.

Among those attending the funeral was Auguste Comte, whom Saint Simon had met in 1817 and who had briefly been his secretary.

Comte's major contribution to the sum of knowledge might be said to have sprung from a single remark by Saint Simon, which informed Comte's view of the world: "The only absolute is that everything is relative." Comte drew on Saint Simon's positivist view of human society to develop his own philosophy, today known as "positivism." He held that human history was divided into three stages: the theological, when men believed in gods and demons; the metaphysical, when descriptions were sought for "forces" in nature; and the scientific. Humanity must therefore apply scientific principles in government.

The only means of assuring order was to place social rule on a positive or scientific basis. There could be no world harmony while some people explained the world in terms of theology and metaphysics.

To this end the primary study of science should be the nature of
47 29 *58* Man and the development of what Comte called Social Physics,[47]
through which laws of behavior could be applied to the study of society. All perception and understanding of the world came from the

brain and the senses and from this Comte deduced that the development of human knowledge was, as Simon had said, a matter of what people knew at any time. This was affected by their historical context. In this sense all knowledge of the world was relative. By 1850 Comte was an examiner at the Paris Polytechnic and known throughout Europe. Such eminent men as John Stuart Mill considered him an intellectual guide and moral mentor.

In the 1860s a physics professor in Prague looked more closely at Comte's concept of the relative nature of perception in a series of experiments, including one in which subjects were placed on a rotating chair with a paper bag over their heads. The experiment revealed that the subjects were able to perceive rotation only during the phases of acceleration and deceleration. When the chair rotated at a constant speed neither movement nor direction could be identified in the absence of external cues. The same phenomenon also occurred during linear motion. The physics professor, Ernst Mach, theorized that these perceptions were controlled by messages to the brain from the liquid in the ear's semicircular canals.

By the time Mach returned to Vienna (where he had qualified), he was famous for his crowded public lectures on subjects such as "The Role of Accident in Invention and Discovery." By this time he had extended his relativist view of perception to all aspects of scientific investigation. In one lecture he stated: "When we say the acceleration of a freely falling body is 9.810 meters per second, we mean the velocity of the body with respect to the center of the earth is 9.810 meters greater when the Earth has performed an additional 864,000th part of its rotation—a fact which itself can be determined only by the Earth's relation to other heavenly bodies. . . . The aim of research is the discovery of the equations which subsist between the elements of phenomena." In this sense, for example, science could only refer events to the passage of the hands of a clock around its face and never to absolute time.

The essence of Mach's thinking rested in a concept he himself never articulated but which was referred to by his most famous disciple as the "Mach Principle." In simple terms it stated that Newton was wrong in postulating absolute space as a frame of reference, because it could not be observed. All mass and motion in the universe could only be relative to the observer's frame of reference, which was relative to other masses and motions, which in turn could only be described in terms of yet further phenomena. So there could be no such thing as an isolated or detached element of experience. As Newton's apple fell, attracted by the Earth, the Earth was also attracted by the apple.

The German physicist who coined the term "Mach Principle" was introduced to Mach's ideas by a friend and once said: "Even those who think of themselves as Mach's opponents hardly know how much of Mach's views they have, as it were, imbibed with their mother's milk." He came into contact with Mach's ideas thanks to a thought experiment he conducted regarding light. At the time light was considered to travel in a "luminiferous," invisible, intangible, elastic medium that permeated the whole of existence and was called "ether." Unfortunately, nobody had succeeded in finding "ether," so the matter of how light traveled was still in question. In the course of thinking about this the German physicist imagined himelf traveling on a beam of light. As Mach had pointed out, this meant that to the traveler the light beam was, in effect, stationary. So since the light was not traveling with respect to the observer, if he held up a mirror, the light carrying his image would not travel to the mirror and therefore he would see nothing in the mirror. It was at this point that a friend referred to Mach's work regarding the nonexistence of absolute space and motion. According to Mach, within the local frame of reference of the light-rider light would still move as it should and reach the mirror to be reflected, no matter how the light might be perceived from outside the light-rider's frame of reference. It was this which led the German physicist, Albert Einstein, to the realization that the speed of light must be the only constant property in the universe.

Another Machian idea about light presented intriguing possibilities. If everything in the universe were affected by everything else, then light should also be affected. Developing this concept, in 1916 Einstein presented an idea that shook classical physics to its foundations. He theorized that light ought to be affected by gravity as if light had mass. Light ought to be affected by a gravitational field. In 1919 came an opportunity to test the theory, thanks to an eclipse of the Sun. On May 29 that year a group of bright stars in the Hyades was expected to form the starfield around the sun (that is, they were directly beyond it in space at the time).

The eclipse track was due to pass just off the coast of West Africa and run through the island of Principe. Photographs were taken of the Hyades starfield several months earlier when the sun was nowhere near them in the sky. Then on May 8, a British expedition headed by Sir Arthur Eddington, Director of the Cambridge Observatory, set out for Principe. On the day of the eclipse sixteen photographic plates were exposed. On July 5, after taking further comparison pictures, the expedition left the island, returning to the Royal Observatory in Greenwich on August 25. Of the sixteen plates taken only two had reliable images. These plates revealed that the light from the stars indeed had been deflected, by 1.75 seconds of arc, as it passed through the sun's gravitational field. Eddington send Einstein a telegram confirming his theory.

One of Einstein's students, noting his unruffled reaction to the news, asked him how he would have felt had the observations not confirmed his predictions. Einstein replied: "I should have been sorry for Eddington. The theory is correct."

CHAPTER 4

An Invisible Object

Not long after Einstein predicted that light was affected by gravity (a prediction confirmed by the eclipse in 1919), the German astronomer Karl Schwartzchild theorized that there could be objects in space so massive they would attract and hold light. The objects would of course be invisible since no light from them could escape to be seen. Such an object would be like a hole into which light "vanished." For this reason the theoretical phenomenon became known as a "black hole."

In 1992 the presence of massive black holes (in galaxy M-87) was confirmed by an instrument named after the astronomer who, like Einstein, revolutionized the modern view of the cosmos. His name was Edwin Hubble, and in the 1920s at the Mount Wilson observatory near Los Angeles, he began a study of extragalactic nebulae, clusters of stars that look like clouds (Latin: *nebulae,* cloud). In attempting to gauge the distance of these nebulae Hubble used a new law described ten years earlier by Harvard astronomer Henrietta Swan Leavitt. She had observed that the brightness of a group of stars named Cepheids waxed and waned regularly. The length of time between the brightest and dimmest state indicated the true brightness of the stars. The magnitude of brightness of a star indicates how far away it is (the brightness diminishes inversely with distance, so a star twice as far away is one quarter as bright). Hubble found varying Cephids in Messier 321, the great star-cloud in the Andromeda constellation. Their magnitude and period of variation revealed them to be 750,000 light years away. This caused astronomers to double the previously accepted size of the universe. In 1929, using the same magnitude test, Hubble observed stars in the Virgo cluster and discovered them to be 250 million light years away.

It was also at this time that he began to measure stellar red shift
known as the Doppler[48] effect, which is caused by the way in which **48** 118 *221*
light waves from a receding source will arrive at the observer's eyes less frequently (the waves having been "stretched out" by the movement of their source). This shift makes the light redder (low-frequency light is redder). Hubble found that the red shift increased with the distance to the stars and from this observation formulated Hubble's Law: At double the distance, the star is receding twice as fast. This astonishing discovery led to a total revision of scientific understanding of the cosmos that was as fundamental as the change in thinking brought about by Copernicus four hundred years earlier. The red shift was observable everywhere and indicated that the entire universe was expanding. By plotting this expansion backward it was possible to arrive at a point in the past when the universe had begun in an immensely dense concentration of matter that had exploded. Hubble's red shift generated the Big Bang theory.

According to this theory the farther out into space telescopes can see the farther back in time they look, since the most distant objects will be those which have been moving away longest. To search out such objects was one of the goals of the astronomical intrument that also confirmed the existence of black holes: the Hubble Space Telescope. The telescope was able to give an unprecedented view of the universe primarily because of its position six hundred kilometers above the Earth and free of the distorting effects of the atmosphere. Hubble was taken to orbit April 25, 1990, by the crew of Shuttle Mission STS-31 and was operated by the Johns Hopkins Space Telescope Science Institute on behalf of NASA and the European Space Agency. Although initially the Space Telescope was due to be returned to Earth at intervals for servicing, not long after it had begun to work NASA decided that maintenance would be better accomplished in orbit.

On both the delivery and later maintenance missions the 121-foot, 4.5-million-pound shuttle, a craft the size of a DC-9 and with a wingspan of 78 feet, maneuvered into the exact position for the telescope's initial release or (on maintenance missions) close enough to the orbiting telescope to grapple it. The delicacy of these maneuvers shows the extreme accuracy with which the shuttle can be flown, thanks to its reaction control system (RCS). The RCS consists of forty-four small nozzles set on the nose and the aft fuselage near the main engines. Thirty-eight of the thrusters each produce 870 pounds of thrust and six of them each produce 25 pounds. The larger engines are designed to be able to do fifty thousand firings, the smaller five hundred thousand. The RCS enables the pilot to position the shuttle to within half a degree and burns nitrogen textroxide and monomethyl hydrazine. Hydrazine is an extremely powerful fuel (first used as a fuel during World War II by the German rocket-powered ME 163 fighter) and provides more bang for the buck (greater specific impulse) than any other fuel except hydrogen. Hydrazine has other less explosive uses: in pharmaceuticals, for corrosion control in water-heating systems, in photography, photocopying, dyes and metal-plating. It is also used in fungicides.

The first fungicide was developed in reaction to the greatest disaster France has ever experienced: the great downy mildew plague that struck French vineyards in 1878. The irony was that the fungus had been introduced on American vine stock brought to France to solve an earlier wine crisis: the great phylloxera epidemic of 1865, which by the time downy mildew appeared had already decimated French wine production.

Toward the end of October 1882 the Bordeaux University professor of botany, Pierre Millardet, was strolling among the vineyards and noticed that some of the vines were a strange greenish-blue color. They were also healthy. Inquiries revealed that it had long been the custom among local growers to spatter their vines with a mixture of copper sulphate and lime to deter would-be thieves. Millardet had been looking for an antifungus chemical with which to coat vine leaves at the point in the year when downy mildew fungus was most vulnerable to chemical attack. By the time of the next harvest Millardet had produced the "Bordeaux Mixture" of copper sulphate, quicklime and water. It was the world's first chemical fungicide.

The wine plagues had halved French output. Since wine accounted for a quarter of the country's agricultural income and provided a living for over six million people, the government was keen to prevent such a disaster from happening again. In 1878 an international quarantine conference took place in Bern, Switzerland, where regulations were agreed on the transfer of plants across frontiers. Unfortunately, the delegates seemed to ignore the fact that phylloxera had been caused by an aphid and that aphids can fly.

Other problems related to the crossing of frontiers were exercising the minds of Swiss conference delegates around the same time. In 1874 twenty-one countries met at a Bern Postal Conference to regulate the delivery of international mail[49] and in particular to establish **49** 78 *149*
agreed charges and categories. Four years later members of the International Postal Union added a Money Orders Agreement. In 1882 an American organization started the first Express Money Order with a tamper-proof order form. The purchaser tore off a row of figures, printed at the side of the form, down to the required amount. This made it impossible for anyone to increase the sum to be cashed at the receiving end. The company, American Express, also guaranteed against loss or forgery.

American Express began in 1850 with the amalgamation of three express services. The impetus to merge had come from the discovery of gold in California two years earlier. The attraction for American Express was the $60 million worth of gold being shipped East in one year alone and the growing demand by the gold miners and businesses in California for faster deliveries from suppliers on the East Coast. To facilitate these deliveries American Express also invented the "Cash On Delivery" system. Then, in 1860, seventy-five thousand Californians signed a petition to Congress demanding a regular and efficient postal service to link them with the families they had left behind. In 1858 gold had been found in Colorado and Kansas. These factors combined to stimulate interest in closing the last gap left in overland communications. It lay between St. Joseph, Missouri, and Sacramento, California.

Fig. 10: *A romantic view of the Pony Express. Most riders did not carry guns, trusting to outride their attackers.*

On April 7, 1860, the first Pony Express rider left St. Joseph carrying mail for California. The route was hazardous in the extreme, crossing nearly two thousand miles of the worst terrain on the continent, inhabited by hostile native Americans and bandits. Riders were recruited for their ability to withstand severe hardship and fatigue. Some were as young as fourteen. Orphans were preferred. One hundred thirty-eight stations lined their route. Every hundred-odd miles came a "home" station where riders rested briefly. At every other home station a new rider took over. At twenty-mile intervals between home stations smaller stations provided fresh horses. Schedules were tight, allowing only two minutes for a change of rider. The Pony Express never went slower than at the gallop and company policy was that whatever the conditions or dangers "the mail must go through." The system lasted less than two years, after which the railroads pushing from East and West joined and made the Pony Express obsolete.

American Express bought the Pony Express for the last few months of its existence and during that time hired an extraordinary rider. He eventually became known as "Buffalo Bill," thanks to his prowess in shooting the animals (sixty-nine in one day) when he was working as a meat-supplier to the Kansas Pacific railroad construction gangs in 1868. William "Buffalo Bill" Cody was a legend before his twenty-eighth birthday, in 1872. That year, having worked for the Pony Express, the railroads, the Army, and as a scout he appeared on the Chicago stage playing the lead role in *Scouts of the Prairies.* After his stage career had begun, periodic returns to spectacular adventures on the prairies added to his reputation. In 1883 he brought the frontier to the city with his Wild West Show. The spectacle acted out a mainly fictitious history of the West, with buffalo hunts, Indian attacks, Pony Express riders, cavalry rescues, sharpshooters, wagon trains, trappers and scouts, and starred Cody as master of ceremonies. By 1913, when Cody went bankrupt, the show had toured a dozen countries and performed before fifty million people in a thousand cities.

The only competition Cody's show faced in America was from another new kind of entertainment, known as "vaudeville." By the end of the nineteenth century the standard vaudeville program consisted of nine ten-minute acts, including a ventriloquist, a dance routine, a comedy and pantomime, a short play and a spectacular finish, often a trapeze artiste. One of the first vaudeville shows had appeared in San Francisco in 1850, billed as a racy "Parisian" entertainment. The link with France is obvious from the genre name, the original French version of which was *Vau de Vire* (River Vire valley). The Vire valley is in Normandy, where the first vaudeville appears to have been invented by a fifteenth-century songwriter named Olivier Basselin. Not much is known about Basselin except that he probably started as a cloth fuller. Most of his musical output consisted of drinking songs with titles like "Fun at the Table," "Drinking Produces Good Verse," and "Let's Have Another Glass!"

The period when Basselin was writing coincided with the gradual loss of Normandy by the English, who had occupied the country for over two hundred years. In 1450 at the Battle of Formigny (the last conflict before the final defeat and withdrawal of the English) Basselin was killed. Formigny was the first engagement in which the French used the new small cannon known as "culverins." Their use was to prove decisive. Traditionally, English tactics were to choose a defensive position, draw up their longbowmen, stick stakes point-up in the ground in front of them and wait for the French to come forward. This would be encouraged with volleys of arrows aimed at galling the mounted French knights into charging. The knights would then be halted by the stakes, and English infantry would move in to finish them off.

At Formigny this plan failed. The French set their culverins to fire along the English front and provoke their archers into action. Some of the archers broke ranks and rushed out to capture the culverins. They were dragging them back to the English side when French reinforcements turned up, saw the break in English lines and attacked. Of the 4,500 English soldiers, 3,774 perished. One further English defeat would mark the end of England's control of Normandy.

The tide had been turning against the English for over twenty years thanks to an illiterate young girl who in 1428 had turned up at the castle of Vaucouleurs in Burgundy demanding audience with Robert de Baudricourt, captain of the town. The girl claimed to have come at the urging of the voices of St. Margaret, St. Catherine and St. Michael. Wearing a "poor, thin" red dress, she announced her sacred mission: to expel the English from France and put the dauphin Charles (heir to the crown) on the throne. Swayed by her piety, De Baudricourt took her to visit Duke Charles of Lorraine, whom she persuaded to lead a less dissolute life. In return for this advice the duke gave her a horse. De Baudricourt added a sword and took her to see the dauphin at Chinon. To test her miraculous powers the dauphin wore ordinary clothes and hid among his courtiers. The girl went straight to him and they talked. Later it was said that she had told him things only he and his confessor knew. Next the clergy questioned her about her heavenly voices, and doctors confirmed that she was also in a rare and mystic state: virginity. When she had passed all the tests the dauphin equipped her with armor, lance and standard and gave her command over his armies, much to the discomfiture of his feuding aristocrats, who resented obeying a commoner.

In 1429 victory over the English at the Battle of Orleans made her name. Henceforth she would be known as Joan of Arc, the Maid of Orleans. In battle after battle she routed the English, and finally on July 17, 1429, she led the dauphin to his coronation in Rheims. Once Charles was safely crowned he regarded Joan's mission as complete and disbanded the army the following September. He reckoned without Joan's promise to remove the English from France. Deprived of the dauphin's army, Joan turned freelance and things started to go wrong. After a series of defeats, on May 24, 1430, Joan rode out of Compiègne to attack the English and their Burgundian allies. After being caught in a pincer movement she returned to the city only to find that the gates were closed and the drawbridge raised. Within hours she was prisoner of the Burgundians.

A year later she was before a court of the Inquisition.[50] The verdict: **50** 104 *197*
"Having weighed the aim, manner and matter of [Joan's] revelations, the quality of her person, and the place and other circumstances, they [her voices] are either lies of the imagination, corrupt and pernicious, or the said apparitions and revelations are conjured up and proceed from malign and diabolical spirits, Belial, Satan, Behemoth." In September 1430 the Burgundians sold her to the English and on May 24, 1431, she was burned to death on a scaffold too high for the executioner to give her the customary *coup de grace*. Left to the flames, Joan died a terrible and lingering death.

The Inquisition that condemned Joan had originally been established two hundred years earlier to combat the heretical Cathars,[51] a **51** 103 *196*
sect that believed in poverty and free love and advocated the renunciation by the church of all material possessions. A young Spanish cleric, Domingo de Guzman, suggested that the only way to combat the growing disaffection (with a bloated church hierarchy that flaunted its wealth and indulged in gambling and fornication) was by

meeting the Cathars on their own ground. His tactic of mendicant preaching failed, and the pope launched the crusade that would by 1209 bring about the massacre of thousands of Cathars. In 1215 Pope Innocent III placed Domingo at the head of a new "Dominican" Order whose only task would be to seek out and combat heresy.

Some time between 1227 and 1233 the problem of heresy had become widespread enough for Pope Gregory IX to take further steps. By the time Joan of Arc was being interrogated the Dominican Inquisition had been set up. With the authority of the Papal Bull *Ad Extirpanda* ("those who must be wiped out"), inquisitors outranked the secular authorities and were empowered to imprison suspects without trial and submit them to torture, to confiscate the property of any suspected person, and in the final analysis to put heretics to death by fire. However, the atrocities of the Inquisition must be viewed in the light of general practices of the day. Torture was common. Punishment for treason in France involved hanging the guilty party until he was almost unconscious, then opening his abdomen and drawing out the entrails, and finally cutting his body into four pieces. Castration was an optional extra. It was also virtually impossible to survive the process of inquisition without confessing. Charges were never specified, and the names of witnesses for the prosecution were withheld. Plaintiffs were asked to imagine which crime might have led to their arrest and then to confess to that crime. They were also promised lighter sentences if they named other heretics. In the fifteenth century the Spanish Inquisition was set up to deal with people easier to identify than scattered groups of heretics: the Jews. The purpose of the new Inquisition was ostensibly to check and confirm the true conversion of Jews. In reality its task was to solve the "Jewish problem" by identifying *conversos* who still secretly professed their faith and dispossessing the richest Jews of their property and position.

The Iberian Jewish population had been in Spain and Portugal since before the eighth century, when the first Muslim invaders had taken the city of Cordova. Muslims regarded Jews as "people of the Book" who worshiped the same God and treated them as a protected minority. Jews were permitted to observe their own religion, were exempt from military service and were allowed their own forms of government. While it was true that in courts of law their testimony carried less weight than that of a Muslim, they could not marry a Muslim, and they were not allowed to bear arms, nonetheless their condition compared favorably with that of Jews in Christian countries. Iberian Jews often reached high public office and were active in trade and finance. Jewish scholars were held in high esteem and played a significant role in the cultural life of the Spanish Muslim states.

After 1100 Muslim Spain began to contract as the Christian "reconquest" moved south, capturing Leon, Asturias, Aragon, Navarre, Catalonia and Castile. The last Muslim states to fall were those of Seville, Saragossa and, in 1492, Granada. Muslim tolerance of the Jews was now to be replaced by Christian repression. Jews were gathered into ghettos, forbidden to leave their homes at night and subjected to taxation that bankrupted them. In 1281 all Jews in Castile had been imprisoned and only freed on payment of a huge ransom. In the fourteenth century the synagogues of Madrid, Burgos, Cordova, Toledo and Barcelona were attacked and destroyed. The Dominicans worked tirelessly to convert Jews to Christianity. Faced with the terrifying alternative thousands of Jews became Christians. Many of these *conversos* rose to positions of power and influence, because after centuries of living in the sophisticated Muslim world they possessed skills such as numeracy and literacy that their new Christian masters lacked.

In 1474 the charismatic Dominican prior of Segovia became confessor to Queen Isabella of Castile, and in 1482 Isabella and her husband, King Ferdinand, installed him as inspector general of the Spanish Inquisition. Tomas de Torquemada's name would become synonymous with the worst excesses of the institution. On January 20, 1492, the capture of Granada, the last Muslim stronghold on the peninsula, brought 781 years of Islamic rule in Spain to an end and doomed Iberian Jews. On March 31 that year Ferdinand and Isabella confirmed the decree expelling Jews from Spain. They were given four months to prepare for their departure and forbidden on pain of death to return to Spain. All their property was to be confiscated;

they were not permitted to take gold, silver or minted coins with them; and anyone giving help, food or shelter to a Jew after July 1, 1492, would be severely punished. The Edict of Expulsion was a disaster for the Jews, most of whom could not realize their worth in the time available. Houses in the ghetto were valueless, and Christians made a fortune by buying Jewish assets at giveaway prices. Best estimates are that 250,000 Jews were driven out.

Iberian Jews scattered throughout Europe, many of them to Eastern Europe, Holland and Portugal, but most returned to the Islamic community that had treated them with understanding and sympathy for so long. The vast majority of these headed for Istanbul, ruled at the time by Sultan Bajazet, who remarked that Ferdinand was a poor statesman, since by expelling the country's most intelligent and industrious members he had impoverished his own kingdom to benefit his rivals.

It was not long before the Jews were performing valuable services in their new home. The year 1520 saw the start of the forty-six-year reign of Bajazet's great-grandson Suleyman the Magnificent, who would take the Ottoman Empire to its cultural apogee and extend its rule over three continents. The Jews were of particular value to Suleyman because their knowledge of European banking systems and their development of a sophisticated system of bills of exchange with other Jewish bankers throughout Europe made them able to transfer funds more efficiently than their Christian and Turkish competitors. Their contacts with Jewish colleagues in the European marketplace, their flair for languages and their ability to keep accounts and write commercial letters and contracts combined to make them critically important to the Turkish exchequer. Jews became indispensable wherever they developed a monopoly, as they did in the trading of sugar, coffee and spices. In 1552 one of these high-placed Jews, Moshe Amon, petitioned the Turkish government to secure the transfer to Istanbul of funds belonging to a Portuguese Jewish banker, Dona Gracia Mendes. This formidable lady had recently renounced her conversion to Christianity and expressed the wish to emigrate to Istanbul together with her nephew Joseph Nasi.

Joseph had spent time in the Mendes family bank in Antwerp and had become a personal friend of the Emperor Charles V. Gracia and Joseph's wealth soon bought them power and influence in Turkey. Gracia became a confidante of Roxelane, Suleyman's favorite wife. By the early 1560s Joseph was fabulously wealthy and was in all but name Turkish foreign minister, in one instance mediating an agreement between Selim (Suleyman's successor) and Charles IX of France. Joseph was extremely ambitious, once styling himself "King of the Jews," although the highest rank conferred on him by the Turks was the dukedom of the Greek island of Naxos.

In 1564, not long after Joseph had gained Suleyman's ear, the sultan's council discussed the possibility of an attack on Malta. The island was the headquarters of the Knights of Malta, a Christian Order Suleyman had expelled from their previous base on the island of Rhodes several decades earlier. Voices at the council were raised against the scheme, arguing that Turkish expansion should be to the north and east, beyond Hungary. In the end the council decided to capture Malta and use it as a base from which to attack Spain, Italy and southern Europe. By this time the Knights of Malta, positioned in the channel between Sicily and North Africa, were indulging in what would best be described as organized piracy, one act of which decided Suleyman on his course of action. The Knights captured a merchant ship belonging to Kustir-Aga, chief eunuch of the sultan's seraglio. The ship carried a cargo of eighty thousand ducats and was loaded with Venetian merchandise in which the principal ladies of the imperial harem had taken shares. Suleyman was therefore under pressure by members of his immediate family to capture Malta, retrieve the goods and free Muslim prisoners (now being used ignominiously as galley slaves on ships fighting the Turks).

On Friday, May 18, 1565, the first Turkish ships were sighted by the Maltese defenders. They were the vanguard of a massive invasion fleet consisting of 181 ships carrying 30,000 men, 6,000 barrels of gunpowder, 1,300 cannonballs, 80,000 rounds of small-arms shot and provisions. Malta's defenders numbered 9,000 fighting men, 5,000 of whom were native islanders. The Knights were superb, highly trained soldiers with a Crusader tradition of courage. So strong was their fear of capture by the Turks that Malta's defenders were prepared to fight to the end. Their leader, Jean Parisot de la Valette, grand master for seven years, had an extraordinary grasp of military tactics and inspired his men by example. He had already spent months preparing the ground, erecting fortifications and razing buildings that offered cover to the attackers. In the long run the tenacity of the defenders and the rifts in the Turkish command, coupled with their total lack of tactical planning, finally led to the attackers' withdrawal and the lifting of the siege.

After the Turks had left almost every major building in Malta was in need of repair or replacement. It was decided by the European powers (which had come too late to the aid of the island) that Malta should be heavily fortified for the key strategic role she was now to play. In December 1565 Francesco Laparelli, a military engineer, arrived to start building. It was decided to erect a fortified capital city to be called Valetta. The new city was the last word in Renaissance design. A rectangular pattern of streets and open spaces was surrounded by walls and bastions. Every house was connected to the town sewers and contained cisterns for collecting rainwater. A uniform style of building was required.

The outstanding new construction in the new city was the hospital, which opened in 1575. The Knights of Malta had begun as a nursing order founded in Jerusalem in 1048 to care for wounded Crusaders, so they were also known as "Knights Hospitallers." The hospital director, known as the grand hospitaller, ranked with the highest officials of the order. The Valetta hospital incorporated many radically new ideas. There were separate wards for different types of disease: one for the treatment of kidney stones, one for venereal and skin diseases, one for hot-bath treatment of syphilis, one for the terminally ill, one for contagious diseases, one for dysentery and most unusual of all, one for mentally ill patients. Another outstanding feature was the provision of single beds, each of which was made up every evening. If necessary sheets were changed several times a day.

Treatment was limited. Anesthesia was administered with a narcotic sponge soaked in belladonna and mandrake and held over the patient's mouth, or by the use of a hammer with which the physician struck a wooden helmet worn by the patient to knock him out. Wounds were usually washed with salt water. Splints and traction were employed. Severed blood vessels were ligatured and wounds in soft tissue were stitched. The corps of physicians operating in the hospital were required to be "learned and experienced" and to swear on oath that they would do their best for the sick. They were also required to base their practice of medicine on that of the "most approved physicians."

The most approved physician of the time was a doctor whose father had been pharmacist to the imperial house of Hapsburg and who himself had been physician to Emperor Charles V. Andreas Vesalius had qualified in Paris and then at Padua, where in 1537 his demonstration of dissection techniques was so impressive he was immediately made professor of surgery at the age of twenty-three. In 1543 he stunned the medical world with a publication entitled *De Humanis Corporis Fabrica* (On the Structure of the Human Body). The book was the first truly modern anatomy text with illustrations taken from life. Vesalius dispensed with the authority of ancient Greek and Roman medical texts and presented what he saw: everything from skeleton to nerves, organs, muscles and skin. Vesalius also described his dissection methods well enough for a practicing reader to follow. The *De Corporis* illustrations were woodcuts carved in pear, sawn with the grain and then rubbed with hot linseed oil to give the surface greater elasticity. The identity of the cutter is unknown but there is little doubt that the extraordinary drawings from which he worked were done in the studio of the great Venetian painter Titian.

Tiziano Vecelli was born of aristocratic parents in northern Italy in 1488 and already at the age of eight showed such a talent for art that he was sent with his brother to be apprentice to a master in Venice. The Most Serene Republic of Venice was the richest seaport in Europe and known for conspicuous consumption. Art offered profitable career opportunities, as Giovanni Bellini, the leading Venetian painter when Titian arrived, had found out. His contract with the government provided him with a life-long pension in return for painting one portrait of each doge as he was elected. It was to the studio of Giovanni's brother Gentile that Titian and his brother were sent.

After a short stay with Gentile Titian left to work for Giorgione da Castelfranco, who had just shocked the Venetian art world with an altarpiece in which he had painted the Madonna standing in a landscape instead of a church. It was Giorgione ("Big George") who for the first time began to use the shading of colors to suggest shape and volume. His innovative idea of using canvas and oils flouted tradition and introduced a realist style that Titian was quick to adopt. In 1526 Titian produced an altarpiece for the Church of the Friars that is reckoned to be the first example of his personal style. It shows a shapely Madonna being lifted to heaven by a host of angels. The radical thing about the painting is that the Madonna looks heavy. The Friars were said to be "shocked" the first time they saw it. Titian's full-blooded realism caught on like wildfire. Titian's women were so realistic because he was a master of flesh tones, the skill that had made the illustrations in Vesalius's book so powerful.

By 1530 Titian was everybody's favorite artist, and he began to paint portraits of the great: the queen of Cyprus, the duke of Ferrara, a Medici cardinal, the king of France. In 1545 he went to Rome and was welcomed by the pope like a prince. His portrait of Paul III was so realistic it is said that when he put it out on a terrace to dry passers-by doffed their hats. The Holy Roman Emperor Charles V thought highly enough of Titian to create him a count palatine and knight of the Golden Spur, with right of entry to the imperial court. This was an honor not normally granted to painters.

Perhaps the most famous of Titian's many portraits of Charles was painted while the emperor was in Germany in 1548. For a number of years Charles had been attempting to find common ground between the Catholic church and the new breakaway German Protestants, who had been led to schism in 1521 by Martin Luther. By 1545 the Protestant princes had formed a league, had refused to attend the upcoming Council of Trent (whose mandate was a sweeping reform of the Catholic church in response to the Protestant movement) and had rejected every attempt at compromise. The emperor had no re-

Fig. 11: *Titian's portrait of Paul III. In 1545 the pope welcomed the painter to Rome "like a Prince."*

sort but to force. In the spring of 1547 at the Battle of Muhlberg, Bavaria, the Protestants were routed, and one of their leaders, the elector of Saxony, was taken prisoner. Titian painted him. Then in nearby Augsburg he painted a portrait of Charles astride his horse in full parade armor, dressed as he had been for the battle.

Charles V's armor had been made in Augsburg because the city was famous for its metalworkers and lay close to a major mining area producing gold, silver and iron. The suit Charles wears in the Muhlberg painting features one of the other things for which Augsburg was famous: a screw. The screw principle had been known since Classical times when it was used in olive presses and to lift water, but it was only in the Renaissance that metal screws were developed, and one of their uses was in armor. Part of the armor, known as a *manteau d'armes* (a rigid shield fixed to the left breast and shoulder and used in tilting) was attached by means of a square-headed bolt tightened with a spanner. The *manteau* was often also a feature of the kind of parade armor Charles wore at Muhlberg.

It is probable that the metal screw was originally developed in Augsburg by jewelers or goldsmiths whose expertise in precious metal work also led to the development of a new method of coinage. At some time around 1550, thanks to the enthusiastic recommendation of the French ambassador to Augsburg, a goldsmith named Max Schwab was invited by Henry II of France to bring one of his new balance presses to Paris. Schwab's press was turned by means of a long transverse wooden handle, weighted at each end. When the handle was turned to screw the press down, the weights on the handle exerted a balanced pressure so that the die held in the press was forced evenly onto the metal and created an unusually sharp impression.

Henry II was keen to reform his coinage because he suffered from a problem common to all heads of state at the time: Bills were settled in cash, of which there was a very limited amount available. In 1530 Charles V had held the sons of the French king Francis I to a 2 million gold ecu ransom. Since there were not enough gold ecus in France to pay the ransom, Francis was obliged to buy foreign coin, melt it down and turn it into ecus. The cost of wars and the mercenaries who fought them frequently obliged kings and princes to borrow money. When Max Schwab arrived in France, Henry II owed the Lyons bankers almost his entire annual income. The situation was complicated by the presence of large amounts of counterfeit coins and the fact that all coinage was easily debased through shaving down. Henry's idea was to create new coins to replace the debased currency. The ecu was to be renamed the "henri" and given a lower gold content. In order better to manage the recoinage (and the currency in general) Henry set up a royal treasury in the Louvre to house money chests to which only Henry and his chancellor held the keys.

One of the major demands on Henry's purse was the expenditures of his wife, Catherine de Médicis, renowned for her love of luxury. On her accession she had doubled the size of the queen's traditional personal retinue to 100. Catherine lived a life of ostentation during the royal progresses round the kingdom. She also personally owned nine chateaux and palaces, one of which (La Tournelle, on the banks of the Seine) she demolished to make way for the Tuileries. She also built a new wing on the Louvre, two new chateaux (Monceau and Chaillot, both near Paris) and added a gallery to the Chateau de Chenonceaux. Her personal apartments were filled with Indian tables, Turkish carpets, hangings of gold and silver cloth, vases of jasper, cabinets inlaid with silver, mother-of-pearl tables, enamel, porcelain, glass, tapestries and hundreds of portraits. It is also said she introduced much of Italian cuisine to France and popularized the idea of restaurants.

Catherine held grand fetes, called *magnificences,* to show off the royal family to diplomats and ambassadors. At one of these occasions she developed one of her famous migraines and retired to take her new medicine. In 1559 the first examples of the newly discovered tobacco plant had arrived in Iberia, and the French ambassador, Jean Nicot, sent Catherine some leaves he grew from seeds given him by a Dutch explorer. Later the great Swedish taxonomist Carl von Linne would coin the term "nicotine" from Nicot's name.

Catherine may have been the first European to take snuff but she was not alone for long. In 1560 Nicot wrote to the cardinal of Lorraine that tobacco had healed an ulcer and eliminated a fistula pronounced incurable by physicians. Over the next twenty years tobacco would be prescribed as an antiseptic, an emetic, and a cure for flatulence, toothache, a heavy cough (inhale smoke deeply), pregnancy pains, halitosis, rabies, gangrene and itching. By 1610 there were few ailments that it was not reputed to cure.

For rulers everywhere the problem with the rapidly growing craving for tobacco was the amount of money leaving their countries to pay for it. Elizabeth of England issued a decree against "misuse" of tobacco. The Turkish sultan Murad IV made snuff-taking a capital offense. Pope Urban III issued an interdict on the taking of snuff in places of worship. None of these measures worked. In London courses in how to smoke were being given by "professors in the art of whiffing." By the mid-seventeenth century the French, Spanish and English governments had awakened to the possibility of turning the situation to fiscal advantage by setting up tobacco-growing colonies in America and state tobacco monopolies at home.

The English colony of Maryland was established almost exclusively for the purpose of producing tobacco. Maryland was one of the three points on the great trade "triangle": English ships would take slaves to the Caribbean, exchange them there for sugar, spices and rum, then deliver these to the American colonists in exchange for tobacco, brought to England. Growing tobacco in the Colony was a night-and-day endeavor, involving thirty-six separate operations, including stirring the ground, transplanting, covering with oak leaves, "hilling," picking off worms, pinching to prevent flowering, cutting, drying, curing, stripping and packing. When the cured tobacco was ready, four-hundred-pound hogsheads were packed tightly with leaves and rolled to the nearest jetty for loading. Some of the roads used for this purpose are still referred to as "rolling" roads. Maryland was an attractive location for tobacco-growing because with hundreds of creeks and inlets around the Chesapeake Bay plantations would rarely be more than a few miles from water.

Early plantations were poor affairs. The majority of Maryland immigrants were single young men, because after the mid-seventeenth century the Barbados plantation owners wanted only black slaves, and New England's foreign exchange shortage limited immigrants to those with skills or capital. Maryland immigrants were granted a piece of land once they had worked off a five-year service agreement. It would then take a further five years to make enough money to get married. Tobacco-growing required few tools: an axe to clear land and a hoe. Most planters lived in one-room frame houses with planks laid across the rafters to make sleeping lofts. The houses were often made of green wood and required frequent repairs, becoming virtually uninhabitable after a decade, when owners simply moved on and built again. This "throwaway house" was a feature of the Maryland countryside that struck most visitors.

The explosive growth of tobacco imports to Europe coincided with a general and rapid increase in international trade as the new European nation-states established colonies and began the exploration of Africa and the East. The opportunity for governments to raise revenue from imports was too good to miss. By the end of the first quarter of the seventeenth century most European countries had updated and extended their excise departments. In 1643 the English government had introduced the first excise duty, initially a levy on home-produced goods such as beer and ale, liquor, cider, soap, meat, salt, leather and cloth. The tax was tremendously unpopular because it was extended beyond the promised single year of collection (and never repealed), and also because a tax on the necessities of life hit the poor hardest. As the number of taxed commodities increased so too did the complexity of administration. The calculation of excise taxes was exacerbated by the fact that different measures were applied to different commodities such as glass, salt, windows, bottles, leather, wood and tobacco. As trade flourished and the marketplace grew more varied the calculations involved in taxation grew ever more complex.

In the early part of the seventeenth century a radically new method of calculation had emerged that would lighten the burden of tax collectors. In 1614 a Scots mathematician named John Napier published a book detailing his new "logarithmic" calculation system. Four years later it had the attention of scientists and mathematicians throughout Europe. In essence Napier made it possible to carry out complex arithmetic calculations through simple addition and subtraction. For example, using a base of 10, 100 can be also written as 10^2 (10×10), and 1,000 as 10^3 ($10 \times 10 \times 10$). Multiplication of the two numbers can be done by adding their logarithms (the small superscript numbers). So $10 \times 1{,}000 = 10^4$. Division involves subtracting the logs: $1{,}000 \div 100$ ($10^3 - 10^2$) $= 10$. Using a base of 2, multiplying 8 (2^3, or $2 \times 2 \times 2$) by 32 (2^5, or $2 \times 2 \times 2 \times 2 \times 2$) means adding their logs ($^3 + {}^5 = {}^8$), and looking up the number for log ^{8} ($2^8 = 256$). Finding the square of 32 means doubling its log (2^5). The result, 2^{10}, is the log of 1,024 (the square of 32). To find square roots, halve the log: the square root of 256 is half its log (2^8), that is, 2^4, which is 16 (the square root of 256). Large and detailed tables were worked out to provide the logs of all numbers, so as to make possible quick calculation of the largest and most complex sums.

The discovery of logarithms soon led to the development of an instrument that would make the work even easier. Some time after 1622 William Oughtred, a Cambridge mathematician, produced a circular brass table carrying two joined pointers that "slid" around scales to perform calculations of any logarithm. Oughtred's slide rule was almost certainly made by Elias Allen, one of the most famous instrument-makers of the day. Allen was one of the first masters of the new London Clockmakers' Company, established by royal charter in 1631 to protect the craft against immigrant foreigners (all clockmakers in London at the time were foreign). The company was given regulatory powers over all clockmaking in England.

One of the foremost London clockmaking firms of the day was run by the Dutch Fromanteel family. In 1658 one of them advertised an amazing new kind of clock: "There is lately a way found out for making Clocks that go exact and keep equaller time than any now made without this Regulator . . . and are not subject to alter by change of weather as others are, and may be made to go a week, or a month, or a year, with once winding up, as well as those that are wound up every day, and keep time as well." Fromanteel was referring to the development by his friend and fellow-Dutchman Christiaan Huy-
52 88 *161* gens,[52] who a year earlier had invented the pendulum clock capable
of measuring time accurately to within ten seconds a day.

Huygens had spent time at the Paris Royal Academy of Science, where his assistant was a Protestant Frenchman named Denis Papin, an extremely skilful mechanic and instrument-maker who also had a degree in medicine. In 1675 Papin turned up in London, possibly as the result of religious persecution in France, took employment with
53 65 *135* Robert Boyle[53] and began a series of vacuum-related experiments. In
53 87 *160* 1679 Boyle's instrument-maker Robert Hooke introduced Papin to
the Royal Society, and Papin showed them his new "digester." This was an apparatus for boiling food under pressure in a closed iron pot fitted with a safety valve.

Papin's digester was to find use as a sterilizer of hospital instruments three hundred years later, following the work of a French sci-
54 3 *27* entist named Louis Pasteur.[54] In 1854 Pasteur took over the Faculty
54 125 *235* of Science at Lille University and began to investigate recent prob-
lems related to sugar beet alcohol, which was inexplicably turning sour. Pasteur discovered that tiny organisms were present in the sour solution. Further experiments with meat juice revealed that similar

Fig. 12: *Louis Pasteur, whose work inspired the Carlsberg brewery to open the first fermentation research lab.*

organisms were present and could be killed by boiling. If a flask of the boiled juices was sealed, a short time after the flask was cracked open again live organisms appeared once more. Pasteur concluded these "germs" came from the air. He then went on to discover the presence of microbes in sour wine. Heating the wine to fifty-five degrees Centigrade would kill the bacteria while not damaging the wine. The same turned out to be true of milk. In a spirited attempt to make French beer as good as its German counterpart Pasteur carried out the same type of research at breweries, with the same results. Pasteur's process became known as "pasteurizing" and was eventually used to sterilize medical instruments in a latter-day version of Papin's digester known as an autoclave.

In the 1870s the German beer-makers, against whose efforts Pasteur's brewery work had originally been aimed, commissioned a locomotive engineer, Carl von Linde,[55] to find ways to keep their beer 55 27 53 cool in summer so as to permit year-round production. Linde responded by developing the first successful compressed-ammonia refrigerator. At around the same time a Scottish engineer named James Harrison was perfecting a similar system in Australia in response to the near-catastrophic food shortage in England, where the rapidly rising population of industrializing towns was fast outstripping the country's ability to feed them.

In the event others than Harrison succeeded in producing techniques for chilling Australian meat, which was delivered by refriger-
56 77 *149* ated ship,[56] and saved England from famine in the 1870s. However, at one point in the previous decade Harrison returned to London and set up a refrigeration plant that would chill paraffin into paraffin wax, ideal for candles. A little later, with the development of novel ways of machine-folding cardboard, paraffin-waxed card containers made possible the introduction of fast foods. In 1906 came the first waxed-card milk containers and the paper cups that would become so familiar as to be invisible at every coffee dispenser in the Western world.

CHAPTER 5

Life Is No Picnic

Instant coffee is a perfect example of modern convenience food. It came into existence in the 1930s when due to good weather Brazil and the other South American coffee-producing countries found themselves with massive coffee-bean surpluses. Various methods of turning this excess into a saleable commodity were tried, and in 1938 Dr. Hans Morgenthaler of the Nestle Company came up with the answer. Coffee was brewed in a gigantic percolator two stories high. The ready brew was then pumped to the top of a tower from which it was sprayed into a drying chamber. As the coffee fell it was blasted with hot air. By the time it reached ground level all the water in the coffee had evaporated, leaving only the fine grounds.

At the time there was no obvious market for the new coffee. Then came World War II and the instant coffee market was instantly created. American quartermasters (and soldiers) were keen to reduce the weight of rations to a minimum so as to keep the size of the ration package small and make the ingredients palatable but not easily spoiled. Eventually, "K" rations became standard throughout the American military. They consisted of three small boxes, each one including a can for each meal (meat, meat and egg, or processed cheese), biscuits, crackers, dextrose tablets, a fruit bar, a chocolate bar, bouillon, lemon juice crystals, sugar tablets, a stick of chewing gum, four cigarettes and instant coffee. The package provided thirty-four hundred calories a day and was easily delivered. This was particularly important in the extremely mobile conditions of World War II when armies were unprecedentedly mechanized. By the end of the war America had over eight million troops in the field, using tanks, trucks and Jeeps. The soldiers were giving quartermasters nightmares because of the speed with which their position changed almost daily. Logistics were extremely complex and large-scale. When fifty thousand American troops went ashore on D-day they were backed up by five hundred thousand quartermasters, medical personnel, ordnance suppliers and signal and transportation corps.

What made the gigantic wartime supply task possible was an extraordinary vehicle specially developed for the purpose: the Jeep. The original specification was for a four-by-four truck, weighing thirteen hundred pounds empty, seating four people including the driver, with minimum ground clearance of six and a half inches, an operating speed range from three to fifty miles per hour and room to carry a 0.30mm machine gun. The manufacturer was to deliver one prototype within forty-nine days and a further seventy vehicles within seventy-five days. The Willys Overland vehicle company won the contract and in the end came in within seven ounces of the all-up specifications because they weighed the paint. The Jeep became the most versatile war machine ever. It was used variously as a command vehicle, weapons carrier, ambulance, cargo carrier, personnel carrier, ammo carrier (with a trailer) and mobile control tower. Above all it was a Jeep that did the final run each day, to deliver rations and supplies to front-line troops. Of the twenty-seven pounds of supplies each American soldier needed to stay fighting no fewer than fifteen pounds of it was gasoline.

Ironically, as the Jeep raced across the battlefields it was using up this precious and limited resource but at the same time riding on tires developed as part of the solution to the fuel shortage. Gasoline was produced by a process developed well before the war and known as "cracking." This involved pumping crude oil through pipes lining a large, white-hot brick furnace, which heated the oil to about eight hundred degrees Centigrade. At this temperature the oil was released into the bottom of a tall steel cylinder known as the "fractionating tower," where all but the heaviest constituents flashed into vapor and

Fig. 13: *The amazing "can-do" World War II Jeep. Names like Peep and Puddle Jumper were also suggested.*

rose up the cylinder. At various levels perforated trays heated to a different specific temperature caused different parts of the rising vapors to condense, the lighter the higher. The last two fractions condensing at the highest levels were kerosene and gasoline. At the top of the tower the final product of this entire process escaped in the form of methane gas.

The Jeep tires were developed because acetylene[57] gas can be de- 57 39 *74*
rived from methane. Acetylene had been around since the late nine- 57 127 *244*
teenth century, when it was discovered almost by accident when
water was dropped on calcium carbide, which gave off the gas. The
Notre Dame University professor of chemistry, Julius Nieuwland,[58] 58 115 *216*
began investigating the properties of acetylene. In 1925 he read a paper to the American Chemical Society about his success in using acetylene to derive an unusual compound: chloroprene. DuPont became interested enough to take the work to the next stage, which

was to polymerize the chloroprene to produce neoprene. Polymerization consists of attaching certain molecules to the ends of other molecules in order to produce gigantically long molecular chains (polymers), which exhibit elasticity and sensitivity to heat. For this reason the generic name for the product of this process is "thermoplastics." The product of this early polymerization, neoprene, was artificial rubber, and the first neoprene tires came off the production lines in 1940 just in time for the Jeep.

The chemist at DuPont who had worked on neoprene was Wallace Carothers. By 1935 Carothers had developed a polymer he named "66" because both of its linked molecules contained six carbon atoms. The "66" was extruded in filaments that could be stretched up to seven times their length when cold and woven into extraordinarily strong fibers that exhibited both elasticity and high tensile strength. The fibers were also sheer, creaseproof, and waterproof. When DuPont released the new fiber onto the market in 1940 in the form of women's stockings, they named it "nylon." When the war was over and nylon could once again be used for civilian purposes it caused a revolution in all kinds of fashion wear.

Nylon stockings were knitted on "Cotton" stocking machines virtually unchanged since 1864, when the stocking-frame machine had reached its final stage of development at the hands of William Cotton, in Loughborough, England. Stocking-frame machines knitted stockings in two pieces, determining the shape of each side of the leg by the number of stitches used in any particular row. The two pieces were then joined with a seam running up the back of the leg.

The early stocking-frame was responsible for one of the most violent episodes in the history of the early Industrial Revolution. In
59 10 *34* 1812 the trade blockade[59] during the war with Napoleon caused a
59 43 *78* slump in commerce because it cut England off from her markets
59 112 *212* abroad. Stagnant trade brought soaring prices and the war brought
high taxation. Average wages fell by one-third. Bad harvests between 1809 and 1811 meant that a four-pound loaf cost almost one-fifth of a worker's weekly income. Profiteering was rife. As the war continued and small manufacturing firms went broke unemployment rose. In the county of Nottingham, the home of stocking manufacture,

piecework weavers usually rented stocking-frames from master hosiers and wove at home (the long windows to catch the best light longest through the day can still be seen in their cottages). As the economic crisis worsened, hosiers began to cut back, reducing wages and increasing rentals.

To all this was added one final, unexpected twist: a change in fashion. Around 1790 people had begun to reject knee breeches and stockings in favor of the new long trousers. Nobody cared about what kind of stockings they wore under their trousers, so manufacturers began to cut stockings out from large pieces woven on ordinary looms. The stocking weavers faced ruin. On March 11, 1811, a weavers' demonstration in the city of Nottingham was broken up by a company of dragoons. The crowd moved on to the nearby village of Arnold, where the rioters broke into cottages and smashed sixty frames. Over the next few weeks the pattern was repeated all over the county. Soon proclamations and pamphlets began to appear, written by the rioters' leader, who called himself General Ned Ludd. It was at this time that the word for machine breaker, "Luddite," entered the language.

The rioters became better organized and began to wear masks and scarves to hide their identity. The movement spread rapidly to the textile manufacturing towns of Yorkshire. Curfews were imposed and 3,000 troops mustered. The country suddenly recalled the riots and violence of the recent French Revolution and the government overreacted. In February 1812 the Home Secretary introduced a bill enacting the death penalty for frame breaking. The same month, in Nottingham, nine men charged with frame breaking (two of them sixteen-year-old boys) were transported to Australia for between seven and fourteen years. The outcry was immediate: Their sentences had been too lenient. As a result, people were transported merely for administering the Luddite oath. In January 1813 at York fourteen men were executed. They were young, hardworking and deeply religious. They sang hymns on the gallows and the crowd joined in.

As the death penalty bill was being debated in Parliament the Luddites found support from an unexpected quarter. A young, unknown

60 30 61 aristocrat[60] rose to give an impassioned speech in the House of
60 132 249 Lords. It was his maiden speech and in it he spoke of men "convicted of the capital crime of poverty. . . . Are not their capital punishments enough in your statutes? Is there not blood enough upon your penal code that more must be forced forth to ascend to Heaven and testify against you? Are these the remedies for a starving and desperate populace? Will the famished wretch who has braved your bayonets be appalled by your gibbets? When death is a relief (and the only relief, it appears, you will afford him) will he be dragooned into tranquillity?"

A few weeks later with the publication of his epic poem *Childe Harold* the young speaker, Lord Byron, was a household name. He said, "I awoke one morning and found myself famous." Five hundred copies of the poem were sold in the first few days, five thousand by the end of the month. The poem had everything: a call to liberty, a hint of libertinism, Eastern mystery, doom-and-gloom forebodings, the tale of a hero outcast for unnamed sins. Women swooned at Byron's feet, seeing the poem as a love letter written for them alone.

At the time of his defense of the Luddites in 1812 Byron had just returned from the journey during which he had written *Childe Harold*. The trip had been financed by a friend who raised a loan on a gambling win. Byron's journey to the mysterious East was in part inspired by the tradition that rich young aristocrats with a passion for culture had for the previous seventy-five years or so been obliged to take the Grand Tour of Europe. Initially in the mid-eighteenth century this had been triggered by growing interest in such archeological discoveries as Pompeii and Herculaneum. By Byron's time the emphasis had shifted, now that the Greek Revival phase of the Romantic Movement was in full swing. The focus was now on the struggle to free Greece from the yoke of the Turk. Byron's was a fact-finding mission, initially to Turkey and then to Greece and Albania, both under Turkish control. During the trip Byron did everything a young Romantic was supposed to: he dressed in Albanian costume, drank wildly, danced Greek dances, fell in love every day (with women and men), sat up all night around campfires plotting revolution, swam the Hellespont and visited Troy.

At one point on the trip Byron was spotted in the Gibraltar garrison library by a Scotsman named John Galt, who had traveled out from England on the same ship. The initial reason for Galt's journey was to set up a Gibraltar branch of the Glasgow textile manufacturing firm Kirkman Finlay. Galt also had a plan to find a way around the Napoleonic War blockade that had caused such problems for the stocking-weavers and virtually ended all import-export in Britain. For a newly industrialized country this was little short of disaster, and British manufacturers were desperate to find ways around the problem.

Galt's mandate from Kirkman Finlay was to find a route into central European markets through the back door. For Galt this meant moving the goods (a hundred bales of cotton) across the Mediterranean into Turkey and then into Europe over the Turkish-Hungarian border. In the long run the plan came adrift when Galt found himself on the Turkish-Hungarian border with the goods, forty-five camels and a no-show contact. Galt was obliged to get rid of the goods at cost to a local Turk. After this he returned to England, married and became a journalist. Success finally came when he published his *Life of Byron* in 1830.

Meanwhile one other side effect of the Anglo-French blockade was causing trouble elsewhere. In answer to wartime needs for more sailors the British navy had for some decades resorted to press-ganging. By 1806 the navy had eight hundred ships and keeping them staffed required up to 150,000 men. Press-ganging usually involved sending raiding parties ashore to find and drag unwilling men off to sea. The practice became an international issue when the British started doing it to American ships. The Americans saw this as a direct violation of their sovereignty, but the British argued that the American merchant marine included thousands of British deserters who had obtained false certificates of American citizenship.

Things came to a head in Norfolk, Virginia, in 1807 when the USS *Chesapeake* was being readied for a Mediterranean cruise. The British consul in Norfolk officially complained to the yard commander, Captain Stephen Decatur,[61] that the ship was carrying four British desert- **61** 12 *34*
ers. Decatur refused to accept this and the *Chesapeake* set sail. At

Fig. 14: *The end of the USS* Chesapeake, *in 1813, off Boston, being boarded by the victorious British ship* Shannon.

3:00 P.M. as the ship was clearing the area and the crew were still busy stowing provisions and cargo, the *Chesapeake* was approached by the British warship *Leopard* with a request that she heave to so that *Leopard* could put aboard dispatches destined for England. Instead of dispatches, however, *Leopard* delivered a demand for the release of the four "deserters." When the *Chesapeake*'s captain, Commodore James Barron, refused, the *Leopard* opened fire, killing three American sailors and wounding eighteen. A British party then boarded the *Chesapeake,* took the four deserters and left. In reaction to this act of violence President Jefferson imposed a blockade on all British trade with America. The blockade would ultimately lead to the War of 1812.

In August 1814, after burning the White House, the Capitol and all other Washington buildings except the Patent Office, the British army attacked Baltimore. Their immediate target was Fort McHenry, which commanded all sea approaches to the city. On the afternoon of September 13 the British arrived and found the fort well-nigh impregnable from land. It was decided to soften up the city with a bombardment from the sea. The British "bomb" ships *Terror, Metero, Aetna, Devastation* and *Volcano* were brought to within two miles of the fort and the bombardment began. It lasted for twenty-five hours (from 6:00 P.M. on the thirteenth until 7:00 A.M. on the fourteenth). Up to one thousand eight hundred shells and Congreve rockets were fired into the fort, the aim being to cause the defenders to panic and desert. During the attack a young American lawyer who had earlier been attempting to arrange the release of an American civilian found himself negotiating on board the British admiral's flagship when news came of the British plan of attack. The American's release was postponed and they were returned to their truce boat, manned by Americans, from which they were able to observe the entire bombardment.

The last view of the fort they had on the evening of the thirteenth was of the American flag pierced by British shells. One of their British guards told them to take a long look at it, since by morning it would no longer be flying once the defenders had deserted the fort. In the gray light of dawn as the bombardment ceased they looked toward the fort. The flag still flew. The attack had failed. The young lawyer, Francis Scott Key, was so overcome by the situation he dashed off a song that would one day become the country's national anthem. He called it "The Star Spangled Banner." The song first appeared as "The Defense of Fort McHenry," published in the *Baltimore Patriot and Evening Advertiser* on September 20, the first day the paper recommenced publication after the war had caused its temporary closure.

The irony was that this most patriotic of all American songs had been written to an English tune titled "To Anacreon in Heaven." At the time the music was already well-known in England, having been written in 1766 by John Stafford Smith, perhaps the first English musicologist, who had risen to prominence in the 1750s when he wrote a number of very popular catches and glees (short, amusing songs). In 1766 the Anacreontic Society was formed in London and Smith wrote the music to a poem with the same title, which had been written by the Society's first President, Ralph Tomlinson.

Society members were well-heeled men who met once every two weeks to have dinner, sing songs, drink and read poetry. The Society primarily existed to write and recite "anacreontics," a relatively obscure form of verse first written by a sixth century B.C.E. Greek writer named Anacreon. Born in 570 B.C.E. at Teos, an Ionian city of Asia Minor, Anacreon flowered in Athens, where he wrote erotic poetry about the pleasures of love and wine. Anacreon's work had been rediscovered in 1554 by a French publisher named Henri Estienne, who came across the poems in a neglected manuscript at the library of the Dutch University of Louvain while he was on one of his usual hunts for Classical manuscripts. Henri was the third in a line of family publishers, who had started business in 1502 when the grandfather, also named Henri, established himself in the Paris University book trade. After he died in 1520, his son Robert began to make a name with Greek editions. Because of religious difficulties (Robert was Protestant) in 1550 he and his family left for Calvinist Geneva where Robert became a Genevan citizen and opened a printing house. It was Robert's eldest son, Henri, who would introduce Anacreon to Europe.

Young Henri had been brought up in a cosmopolitan printing house atmosphere, where a staff of ten different nationalities each dealt with texts in their mother tongue. Much of the time Henri was obliged to speak Latin, the *lingua franca* among scholars. By the time he was fourteen, Henri already also knew Greek. Two years later he left for Italy on the pursuit that occupied all intellectuals at the time, the search for classical Greek and Latin manuscripts. The late sixteenth century was the time of the great discoveries of "lost" ancient manuscripts in almost every field of classical scholarship: botany, pneumatics, physics, chemistry, medicine, geography, philosophy, metallurgy and literature. Continual discoveries of ancient classical manuscripts amazed and enlightened Renaissance thinkers. The manuscripts naturally required glosses—analyses of what the ancients had meant by their use of terms. This exercise attracted scholars in every discipline. Each time a new manuscript became available for analysis, editors would first collate all earlier manuscripts already written on the same subject and produce a definitive version. This process of analysis and synthesis laid the basis for what would become the scientific revolution of the early seventeenth century. The editorial intellectual fallout also generated new scientific disciplines and put old ones on a firmer footing.

It was Henri Estienne's son-in-law Isaac Casaubon who would lead the field in these endeavors, because it was he who turned the art of glossing into an analytical technique that could be applied to any text regardless of subject matter. Casaubon was another Genevan Greek scholar, born nine years after Estienne's father had arrived in Geneva, whose Protestant family had also fled persecution in France. In 1578, aged nineteen, Casaubon was sent to the Geneva Academy that Calvin had founded and by the age of twenty-three was appointed professor of Greek. By 1591 his reputation as a Greek scholar was established throughout Europe. In 1586 Casaubon married Estienne's daughter Florence who then bore him eighteen children.

Casaubon importuned anybody he could find to collect manuscripts for him. In many cases he received copies of originals, penned for him by traveling friends and colleagues. When somebody died and his collection was being sold Casaubon would make sure one of his contacts was there to bid. Sometimes publishers would send him copies of their new editions. Often, at a time before the invention of the publisher's catalogue, the only way to find out what was coming onto the market was to go as Casaubon did to the twice-yearly book fairs in Frankfurt.

In 1596, irritated by the niggardly Genevan authorities who refused to provide him with either books or adequate recompense, Casaubon accepted an invitation to teach at the University of Montpellier in France, where at last his worth was recognized. By now he was the leading Greek scholar on the continent, except for another Huguenot refugee from France who had settled in Holland at the University of Leiden. His name was Joseph Justus Scaliger, and in 1593 when the thirty-four-year-old Casaubon first made contact with him Scaliger was already a grand old man of scholarship at the age of fifty-three. The friendship began when Casaubon plucked up courage to send a letter of greeting. He received no reply. Then an

English friend wrote to say Scaliger had recently read Casaubon's new edition of a classic text and was much impressed. Two letters then arrived direct from the great man (who was not noted for his open and friendly disposition), and a correspondence began which lasted until Scaliger's death in 1609. In that time Casaubon wrote Scaliger over twelve hundred letters. Later Casaubon would say of the man he regarded as a mentor that he was "a man who, by the indefatigable devotion of a stupendous genius to the acquisition of knowledge has garnered up vast stores of uncommon lore. And his memory had such a happy readiness that whenever the occasion called for it, whether it were in conversation or whether he were consulted by letter, he was ready to bestow with lavish hand what had been gathered by him in the sweat of his brow."

Like many of his contemporaries Scaliger was a traveling scholar who moved from place to place (including at one point the court of
62 106 *204* Mary Queen of Scots)[62] to avoid wars and persecution. Like many of his contemporaries he settled finally in Holland, the most tolerant country in seventeenth-century Europe. By this time Scaliger had already produced his greatest contribution to scholarship, in the form of a new chronological system that became known as the "Julian period." The work was intended to address the problem exacerbated by each newly discovered ancient manuscript. The text would often contain references to dates that did little to help place the manuscript historically, since the dates tended to refer to local events such as a battle, or a siege, or the birth of a child, or the appearance of some heavenly sign. Most often dates related in some way to the in-house chronology of the institution in which the manuscript had been originally copied. These varied dating systems bedeviled textual analysis, since it was often essential to know the date of a manuscript to know whether it had been written before or after others.

Scaliger decided to produce a foolproof chronological system in which all events could be dated with complete accuracy. To do so he took as a base three time-cycles: the twenty-eight-year solar cycle (ending each time the days of the week in the Julian calendar repeated on the same date), the nineteen-year lunar cycle (ending when the phases of the moon recurred on the same days of the week), and the fifteen-year period of the Diocletian tax census. Calculating backward, Scaliger was able to say that the first time all three cycles had started on the same day was in the year 4713 B.C.E., so he called this "Year One" of his new system. Starting at this point, and running all three cycles simultaneously, any historical date could be identified by three reference points. Twenty-nine years after Year One, for instance (4684 B.C.E.) would be Year 2 in the latest solar cycle, Year 10 in the latest lunar cycle, and Year 14 in the latest taxation cycle. Scaliger would have referred to 4684 B.C.E. as "2:10:14." Since the overall triple cycle would only repeat every 28 × 19 × 15 (7,980) years Scaliger reckoned his system would serve its purpose for the foreseeable future.

Unfortunately for Scaliger, in 1582, a few months before the publication of the great work (which he called *A Treatise on the Correction of Chronology*) Pope Gregory decreed the switch from the old Julian calendar (on which Scaliger had based all his work) to the new Gregorian calendar, thus rendering Scaliger's work virtually useless at a stroke.

In the last few years of Scaliger's life at Leiden he had established himself as a leading humanist scholar and became a welcome guest at the court of the Dutch ruler, Prince Maurice of Nassau, who from time to time would ask Scaliger to handle problems like the translation of an Arabic letter the prince had received from a Muslim king. Scaliger and other scholars found refuge in Holland primarily because the country had thrown off the yoke of Spanish rule and, with it, severe constraints on intellectual freedom. The leading player in the gradual removal of the Spaniards from Holland[63] was Prince **63** 122 *230*
Maurice himself, who gradually recovered all Dutch territory north of the Rhine and the Meuse. In 1596 the independence of the new Dutch republic was recognized by England and France.

Maurice's other great success was his reorganization of the army. He always claimed that his military reforms sprang from his interest in ancient Roman military tactics, but much more likely is the way Maurice's thinking was influenced by developments in battlefield technology. For the previous three hundred years war had been dominated by cavalry and pikemen. Large formations of as many as three thousand men (known as phalanxes or squares) carrying twelve-foot-long pikes could either form impregnable defensive "hedgehog" formations with pikes lowered all round, or crush opposition by sheer weight of numbers. In the years before Maurice's reforms these pikemen had also begun to act as a defensive wall around the new, slow-firing muskets. In the 1590s Maurice realized there was a way to improve the rate of musket fire by drawing musketeers up in ten ranks so that the front rank could fire and then retire to the rear to reload while the second rank fired and then did the same, giving way to the third rank and so on. In this way an almost continuous volley of fire was maintained. However, this arrangement also exposed large numbers of men to enemy fire and made discipline and concerted movement more important than ever before. This in turn required elements of standardization in both movement and weaponry.

In 1599 Maurice equipped the entire army with weapons of the same size and caliber, and his brother John began work on a new tool for training: the drill manual. John analyzed all the separate movements associated with the use of pike and musket and gave them numbers. In 1607 a book was published by Jacob de Gheyn and soon translated into English (as *The Book of Arms*), French, Danish and German. It contained detailed illustrations for the sequence of movements to be followed by soldiers, dividing up the use of pikes into thirty-two different positions and the loading and firing of muskets into forty-two separate movements. In 1616 John opened a military academy at Siegen where young gentlemen were trained in the use of weapons, armor, maps and terrain models. Several training manuals were also produced, all based explicitly on Dutch military practice. The presence of numerous foreign units, including English, French, Scottish and German troops, ensured that Maurice's ideas would soon spread. Unfortunately, Maurice's new tactics were never tested in a full-scale pitched battle.

It was only a quarter of a century after their introduction that the reforms were to show their true potential at the hands of the Swedish king Gustavus Adolphus. Thanks to constant drilling and practice Gustavus was able to increase his musketeers' rate of fire enough to

Fig. 15: *From Jacob de Gheyn's "weapon-handling" manual, four illustrations for loading the musket before discharge.*

maintain volleys from only six ranks. Firepower was also greatly enhanced by Gustavus's introduction of standard-caliber field guns, some even supplied with cartridges and capable of a rate of fire of twenty rounds an hour, which was not much slower than muskets. Gustavus's critical improvement on Dutch practice was to make his musketeers advance ten paces before firing, then stand where they were to reload, while the following ranks passed them by to fire and so on. In this way not only was a constant volley maintained but it also advanced toward the enemy. The superiority of the technique was devastatingly demonstrated at the Battle of Breitenfeld, just outside Leipzig, on September 17, 1631, when Gustavus beat a Catholic imperial army that outnumbered him three to two. The victory briefly made Sweden a world power. Small wonder that Gustavus earned the title: "Captain of Kings and King of Captains."

When Gustavus died in battle in 1632, his heir was a six-year-old girl, Christina. On February 1, 1633, Christina was proclaimed king of Sweden (all Swedish monarchs were kings; only the wife of a monarch was a queen), and for the next thirteen years the country was run by a regent, Axel Oxenstierna. In 1644 Christina came of age and set about the greatest achievement of her brief reign: bringing to an end the bloody war against Denmark. Christina was extraordinarily gifted and fluent in German, Greek, Latin, French, Spanish and Italian. She was obsessed by her "ugliness," never taking longer than fifteen minutes to wash and dress, throwing on whatever clothes happened to be at hand, heedless of her appearance. Visitors to the Royal Palace in Stockholm were struck by her habit of wearing men's shoes. She also had a passion for culture and learning, and because of her ready wit became known as the "Minerva of the North."

After being formally crowned king in 1650, her abdication four years later stunned Europe and especially her own countrymen. She chose as successor her cousin Charles. He was crowned on June 6, 1654, the day she abdicated. That night Christina left the country dressed as a man, with her hair cut short. On December 23 she formally entered Rome, was greeted by cardinals and senators, processed to St. Peter's and (to the amazement of Europe) on Christmas Day was accepted into the Catholic church by the pope himself.

Christina settled into Rome and was soon at the heart of its cultural life. Her palace contained the greatest collection of Venetian paintings ever assembled. She founded the Arcadia Academy for philosophy and literature. It was at her urging that the first public opera house was opened. She sponsored Alessandro Scarlatti and Angelo Corelli. She amassed an enormous collection of books and manuscripts and worked to protect the Jewish community in the city. She also (it is suspected) had a long-term affair with Cardinal Decio Azzolino, a leading figure in Vatican politics.

It was while she was still in Sweden that she made what was perhaps her greatest mark on the intellectual life of Europe. It was her habit to surround herself with brilliant and eminent foreign artists, scholars and musicians, and in 1649 she invited René Descartes to come and be her philosopher-in-residence. The enterprise turned out to be a disaster. Descartes was asked to write verses to accompany theatrical productions, to take part in ballets, to write a libretto and on the occasions when he gave lessons to Christina she preferred to discuss literature rather than philosophy. The worst problem was the timing of the lessons, which Christina insisted take place in her library at 5:00 A.M. Six months after arriving in Stockholm, struggling through the bitter Swedish winter dawns, on February 11, 1650, Descartes contracted pneumonia and died.

Descartes was another of the intellectual exiles who, like Scaliger, had found refuge with the Dutch. For a brief period in 1618 he joined Maurice's army as a military engineer. Before this, after graduation he had become more and more obsessed by the fact that his years of study at university had provided him with knowledge that was useless and uncertain. The study of the classics, he wrote, involved "those who are too interested in things which occurred in past centuries . . . [and who] are often remarkably ignorant of what is going on today." Literature "makes us imagine a number of events as possible which are really impossible . . . those who regulate their behavior by the examples they find in books are apt to fall into the extravagances of the knights of romances." The study of philosophy by the greatest minds had not "produced anything which is not in dispute and consequently doubtful and uncertain."

About knowledge in general Descartes concluded: "When I noticed how many different opinions learned men may hold on the same subject, despite the fact that no more than one of them can ever be right, I resolved to consider almost as false any opinion which was merely plausible. . . . From my childhood I lived in a world of books . . . taught that by their help I could gain a clear and assured understanding of everything useful in life. . . . But as soon as I had finished the course of studies which usually admits one to the ranks of the learned . . . I found myself so saddled with doubts and errors that I seemed to have gained nothing in trying to educate myself unless it was to discover more and more fully how ignorant I was."

In 1637 these thoughts led to the great *Discourse on Method* that was to lay the foundations of reductionist thought and prepare for the rigorous disciplines of modern science. Descartes's methodical doubt set the ground rules for a system of intellectual analysis that would render more reliable the discoveries of thought, prescribing as it did a systematic process for thinking things through. The route to certainty was to begin by doubting everything and then to take as axiomatic whatever survived this exercise. Descartes regarded the only certainty to be the existence of the doubting mind. He summed this view up in his axiom: *cogito, ergo sum* (I think, therefore I exist). Attacking the speculative and metaphysical nature of the scholastic modes of thought that had preceded him, Descartes turned his attention in 1633 to the nature of biological life in *Treatise on Man*. He approached the workings of the body and brain mechanistically, describing the ten principal functions of the body (digestion, circulation, growth, respiration, sleep, sensation, imagination, memory, appetites and movement) in terms of mechanical systems.

It is possible that Descartes was influenced in his thinking by the latest high-tech gadgetry in the gardens of the royal château at St. Germain-en-Laye, the principal royal residence at the time. In 1598 Tommaso Francini, a Florentine architect and mechanician, arrived to embellish a series of terraces in the gardens with grottoes and fountains. Using water from the Seine, Francini devised an extraordinary system of waterworks whose main feature was a giant fountain. Water from the fountain then descended into a large number of reservoirs in the galleries lower down the slope. Secondary tubes from these reservoirs supplied other fountains and drove water-powered automata all around the gardens. In one area there was an animated dragon, an organ player and a Neptune. Grottoes of Hercules, Perseus and Andromeda, and Orpheus featured these mythological figures, which, when visitors stepped on hidden plates in the floors, performed complex movements. Perseus, for instance, descended from the ceiling and used his sword to slay a dragon that had arisen from the water basin. In another grotto Bacchus sat drinking from a barrel.

Descartes referred to these extravaganzas in his writings on the brain: "Truly one can well compare the nerves of the machine that I am describing to the tubes of the mechanisms of these fountains, its muscles and tendons to divers other engines and springs which serve to move these mechanisms, its animal spirits to the water which drives them, of which the heart is the spring and the brain's cavities the water main. Moreover, breathing and other such actions which are ordinary and natural to it, and which depend on the flow of the spirits, are like the movements of a clock or mill which the ordinary flow of water can render continuous."

Descartes thought the cerebrospinal fluid in the internal cavities of the brain worked like the garden water supply, flowing down through nerves to power the motion of muscles. This engineering view of the brain excited an English doctor named Thomas Willis, who in 1664 was the most successful physician in Oxford. He was also one of a group of intellectuals, including among others John
Evelyn,[64] John Wilkins and Robert Boyle,[65] who met regularly at **64** 111 *210*
Wadham College to discuss the latest scientific discoveries and in **65** 53 *114*
particular the various experiments and applications triggered by **65** 87 *160*
Evangelista Torricelli's recent discovery of the vacuum. In 1664 Willis broke new ground with his book *The Anatomy of the Brain.* Basing his approach on Descartes's mechanistic view, Willis described in detail the central, peripheral and autonomic parts of the nervous system. Using pathological and clinical observation, Willis divided the brain into different functional areas, postulating that the cerebrum was the seat of thought while the cerebellum controlled in-

voluntary motor function. In many cases Willis's dissection methods were far from exact but his empirical methods were a considerable improvement over earlier, purely philosophical speculation.

To describe his activities Willis coined the word "neurology." While some of his ideas did not stand the test of time, the book was the most complete and accurate account of the nervous system so far and stimulated much further research. One reason for the book's extraordinary success (it became the standard textbook on the subject for 150 years) was its illustrations. These were the first properly modern views of the brain, beautifully and precisely drawn by one of Willis's colleagues at the Wadham College meetings, Christopher Wren.

Two years after the publication of Willis's book came an event that was to shape Wren's life. At the time he was Savilian Professor of Astronomy at Oxford and a classic polymath. His list of "things to think about," which he made shortly after graduating, included "hypothesis of the moon in solid; to find whether the earth moves; the weather wheel; a perspective box for surveys; several new ways of graving and etching; to weave many ribbons at once with only turning a wheel; improvements in the arts of husbandry; divers new engines for raising of water; a pavement harder, fairer and cheaper than marble; to grind glasses; a way of embroidery for beds cheap and fair; pneumatic engines; new ways of printing; new designs tending to strength, convenience and beauty in building; divers new musical instruments; a speaking organ; new ways of sailing; probable ways for making fresh water at sea; the best way for reckoning longitude and observing at sea; fabrick for a vessel at war; to build in the sea forts, moles etc.; inventions for better making and fortifying havens, for clearing sands and to sound at sea; to stay long under water; submarine navigation; easier ways of whale-fishing; new cyphers; to pierce a rock in mining; to purge or vomit or alter the mass by injections into the blood; anatomical experiments; to measure the height of a mountain only by journeying over it; a compass to play in a coach or the hand of a rider; way of rowing; to perfect coaches for ease, strength and lightness."

This unusual mind was provided with a single task large enough to engage it in 1666 with the Great Fire of London. It broke out on the night of Sunday, September 2 and burned for five days before being extinguished. The fire destroyed about 80 percent of the City of London, including the Customs House, the Guildhall, the Exchange, six prisons, eighty-seven churches and the cathedral of Old St. Paul's. The situation after the fire was particularly dire, since there was no insurance to recompense any who had lost their homes and businesses. Such people desperately needed new homes, new storehouses, new offices and a new Exchange without delay. Within days of the end of the fire Wren was one of those who offered King Charles II a plan for a new London. On September 11 the king announced that the city was to be rebuilt in brick and stone with wider streets across which a fire could not jump. He appointed six commissioners to oversee the work. One of these was Wren, who spent time on several small commissions and then in 1669 at the age of thirty-six was appointed royal surveyor and offered the commission to rebuild the cathedral. Thirty-nine years later, in October 1708, the final stone was added to the lantern on top of the great dome and the work was complete. By this time Wren had also been busy elsewhere and the London skyline was dotted with evidence of his architectural skill. Wren built a total of fifty-two churches, twenty-eight of which survive today.

In 1679 Wren became a member of the council of the Hudson's Bay Company and for five years was an active director and stockholder. Since the beginning of the seventeenth century, when the first cargoes of tea and porcelain started arriving from China and Japan on
the ships of the Dutch East India Company,[66] more and more money **66** 44 *80*
had been invested in these speculative journeys of trade and explo- **66** 109 *208*
ration. England soon followed the Dutch example. The Eastland Company had a monopoly of Baltic trade. The Muscovy Company traded as far as Persia. The Turkey Company went to Basra, and one of their ships even reached Malacca. Most sea captains returning from these journeys were interviewed by the Reverend Richard Hakluyt, lecturer in geography at Oxford, whose collected data came to-

gether in his *Principall Navigations, Voyages and Discoveries of the English Nation,* which became required reading for those thinking to set up in the overseas trade business.

As the companies expanded their activities it became clear that the navy was unable to protect each single venturer ship, so they began to band together to organize armed fleets of their own both for protection and to help keep their prices competitive. The stock of each venturer was joined with that of the others and thus the cooperatives became known as "joint-stock companies." By the time Wren was a director of the Hudson's Bay Company the joint-stock market was a thriving concern, attracting investors who had nothing else to do with their money but bury it or buy land. At the start of the eighteenth century the urgent need was for a banking system capable of supporting and facilitating the new stock market.

In 1716 the first joint-stock bank, established to attract investment funding, was set up in Paris. It was the brainchild of John Law, a Scottish gambler and financier. Law's life was a series of extraordinary ups and downs. At some time between 1694 and 1704 he spent two years in Holland acquainting himself with the Dutch banking system. He then moved to Genoa and Venice, gambling and becoming rich. In 1703 he wrote a proposal for the introduction of paper money in Scotland (at the time paper money was in circulation only in Sweden, Genoa, Venice, Holland and England). Law's argument was that paper was more convenient than specie, the scarcity of which was seriously hampering trade. Paper backed by arable land values could be issued in sufficient quantity to release large amounts of money, which in turn would boost economic growth. The Scots, more concerned with the impending Union with England four years later, turned the idea down.

Over a number of years Law made efforts to persuade the French to take up the idea, and in 1716 he succeeded. The Banque Générale was the first private bank, John Law was its first managing director, and it issued paper money. The new money was backed by specie at a fixed rate, and it soon began to be preferred to the fluctuating value of (often counterfeit) French coinage. In 1717 a decree made legal the payment of taxes in paper money, Law began to issue loans at fa-

vorable rates and the nearly bankrupt French economy revived. Law then introduced a scheme for revitalizing French industry with a trading company that would have the monopoly of trade with French Louisiana.[67] At the time this included the entire territory 67 11 34
drained by the Mississippi, Ohio and Missouri rivers.

In 1717 Law's trading company became known as the Company of the West. Law's prospectus painted a glowing picture of hardworking, welcoming natives, as well as of mountains of emeralds and gold that the locals would exchange for knives and mirrors. In 1719 Law proposed an extension of the company to include trading monopolies in Africa, Asia, India and China. This huge corporation would effectively trade with the entire world and generate enormous profits. In 1719 the Company of the West was renamed the Company of the Indes, and speculation in Company shares reached frenzied proportions. Stock worth 1,000 livres in July was worth 6,000 in September. News spread of individuals making more than 1,000 percent profit overnight.

Then came Law's greatest coup: The company was granted the right to collect French taxes and to raise money to liquidate the national debt. For this purpose 300,000 new shares were to be sold. It was now that the first cracks in the edifice began to appear.

The Banque Générale had been so successful it was taken over by the regent, who began to print money at an alarming rate, showering courtiers with extravagant salaries. In future paper money was no longer to be secured by gold, but to be redeemed against the currency of the time. Confidence in the notes began to evaporate. Louisiana stock was grossly overvalued. During the bitter winter of 1719–20 inflation exploded, accompanied by a rapid rise in prices. Devaluation was announced in May. The end had come and the bubble burst. The economy was crippled. Law was exiled and dispossessed of all he owned, including the land on which the Champs Elyses stands today.

Over the following three decades France lost ground to England as the extravagance of the French court, the chaotic state of national finances and the costs of war continued to sap the country's strength. By 1760, the idea was growing that France could only recover at the expense of Great Britain. This anti-British faction was led by Joseph Paris-Duvernay, another financier brought in to stop the rot. In 1760, at the age of seventy-six, Paris-Duvernay met a twenty-five-year-old clockmaker who was radically to change the fortunes of Europe and America. He was Caron de Beaumarchais and at the time he was in residence in Versailles, making watches for the royal family and acting as musical director to the four young princesses. Beaumarchais was also a budding playwright, who would eventually write *The Marriage of Figaro* and *The Barber of Seville.*

In 1776 the king asked Beaumarchais to become a secret agent and go to London to suppress a lampoon being circulated about the king's mistress. In London Beaumarchais became convinced that the British wanted an easy way out of their entanglement with the American colonies and that the simplest way for France to weaken British supremacy would be by covert support for the Americans in the coming War of Independence. In the summer of 1776 a fake company was set up with the blessing of the French government and Beaumarchais began to channel massive amounts of money, arms and

Fig. 16: *Revolutionary Americans tear down the statue of King George III in New York.*

French "military advisers" to the American rebels. After American independence, when the financial effects of this latest adventure hit
the French economy, a Swiss banker named Jacques Necker[68] was **68** 96 *177*
called in to help sort out the mess. In a series of famous "accounts" Necker persuaded Louis XVI that the economy was sound and that the treasury was in the black. The truth was far different. From 1776 to 1786 the state borrowed 1,250 million francs and ran an annual deficit of 115 million. The economy lurched toward disaster. Necker was fired and recalled three times. The absurdity of his stance was greeted each time by increasing invective. Finally in 1789 Necker resigned and returned to Switzerland, leaving his daughter in Paris to enjoy the Revolution that followed.

Germaine Necker had been married three years earlier to the Swedish ambassador to Paris, Eric Magnus Baron of Staël-Holstein. She was now known as Madame de Staël, and her dazzling conversation made her salon the cynosure of French intellectual life. Each morning she would greet her guests in her bedroom, diaphanously dressed and flaunting her libertarian opinions. By 1802 Napoleon was in power and these opinions were less than welcome. De Staël's writing (*On Literature Considered in Its Relations with Social Institutions* and a novel, *Delphine*) had by now made her name all over Europe. That year she traveled to Weimar, Germany, where she was greeted with delight by the grand duke and his family. She met the cream of the new German intelligentsia, including Schiller and
Goethe,[69] though the latter tried to keep out of her way, saying: "She **69** 22 *47*
insists on explaining everything, understanding everything, measur- **69** 82 *153*
ing everything. She admits of no darkness; nothing incommensurable; and where her torch throws no light, there nothing can exist."

It was in Weimar while preparing her book *On Germany,* which would make her a household name among Romantics and establish German culture and the new Romantic movement stirring in Weimar, that De Staël met and captivated August von Schlegel, the foremost theoretician of the new movement. He led her through the Romantic maze and then fell in love with her. On April 18, 1804, when De Staël received the news of her father's illness and was obliged to leave immediately for Switzerland, Schlegel made an instant decision to leave with her and spent the rest of his life as her lapdog. Whether or not he expected the affair to turn passionate it never did. De Staël had other lovers and Schlegel was obliged to resign himself to a platonic relationship.

Earlier, in 1803, Schlegel's marriage had been dissolved and his
70 94 *174* wife had married his friend and colleague Friedrich von Schelling.[70]
Schelling had spent time in Leipzig, tutoring and studying the natural sciences. Out of this came his grand concept, *naturphilosophie,* which was to inform the Romantic movement and much of science for the next forty years. In essence, Schelling's view was that Nature's constant power to evolve and transcend itself, to rise from lower to higher levels and forms, revealed a clear teleological pattern. The overall harmony of Nature was expressed in the way apparent opposites were reconciled. As an example Schelling chose the magnet, in which opposite poles attracted. The same balance could also be found in the interaction of acid and alkali, electricity and magnetism.

71 22 *47* One of Schelling's colleagues, Johann Wolfgang von Goethe,[71]
sought similar purpose and design in embryology. So did other *naturphilosophen*, one of whom was an Estonian professor of anatomy named Karl von Baer, whose aim was to find the unity in Nature expressed in the developmental stages of all organisms. To this end he became expert in the embryonic growth of chicks from the moment of conception to hatching. In 1828 von Baer published a work on the sequential development of the chick embryo, detailing the various stages through which the embryo passed. He discovered the way in which cellular growth proceeded from the general to the particular in the way that the early, homogeneous mass of chick cells separated and then specialized into wings, eyes, beak, legs, the internal organs and so on. Most important of all, von Baer theorized that embryonic development went through different stages in lower and higher life forms. A higher form might pass through early developmental stages that were equivalent to the fully developed forms of lower organisms. This idea was to have the most profound effect on the biological sciences in the latter half of the nineteenth century.

In 1846 HMS *Rattlesnake* left Britain on a voyage of exploration to the South Seas carrying a young naturalist who would take the work of von Baer and others to the next stage. His name was Thomas Huxley, and in 1849 in the waters of Eastern Australia he began a study of jellyfish. When he published his results he noted that the Medusa, with its inner and outer membranes, seemed to be constructed on the same plan as the two primary layers von Baer had seen in the early development of vertebrate embryos. These similarities between the early stages of a higher form and the adult stages of a lower form confirmed von Baer's theory.

Once back in England Huxley would speak in defense of a man who took the implications of these embryonic relationships to their logical conclusion in a great theory of development that sought to describe the processes by which the massively heterogeneous, specialized life forms on the planet had emerged through the mechanisms described in part by von Baer and Huxley. According to the new theory, organisms either specialized to fit their habitat or died.
The proponent of the sober thought that life is no picnic was Charles **72** 4 *27*
Darwin.[72] **72** 74 *145*

CHAPTER 6

Elementary Stuff

There is a respectable body of opinion that holds that the theory of evolution ought not to be known as Darwin's but as Wallace's. Alfred Russel Wallace was a self-educated surveyor, apprentice clockmaker, teacher and beetle-collector who left school at fourteen, became a railway surveyor and developed an interest in geology. His self-improvement took the form of reading the explorer Alexander von
73 117 *220* Humboldt's[73] *Personal Narrative of Travels to the Equinoctal Regions of America,* the works of geologist Charles Lyell, the natural history of Robert Chambers, and Malthus's *Essay on Population.* Lyell's theory postulated an extremely ancient Earth, Chambers's *Vestiges of Creation* suggested that animal species were descended from other animal species, and Malthus's work put forward the notion that populations expanded at such a rate vis-à-vis the availability of food as to make survival a struggle.

Such were the ideas that triggered a wanderlust in Wallace, especially after he had met and been inspired by an entomology enthusiast named Henry Bates. Wallace and Bates would spend many afternoons on field trips collecting beetles. Afterward they would exchange letters about each other's beetle-collecting activities. Wallace saved one hundred pounds and used it to finance a trip to the Amazon with Bates in 1847. There they separated again, Bates taking the Upper Amazon and Wallace the Rio Negro. Five years later, returning with twenty cases of specimens, Wallace's ship caught fire and he lost everything. He had, however, developed an idea that would color his work from then on.

He had noticed that from one part of the forest to another there were major changes in animal and insect life. These differences were manifest even on either side of a wide river. It became clear to Wallace that any species collection ought to be accompanied by notes of a creature's habitat. In 1854 Wallace left England again and spent the next eight years in the Malay archipelago, traveling fourteen thousand miles and building a collection of more than 125,000 different species, the greater part of which was beetles. The collection included no fewer than nine hundred species of longhorn beetle and two hundred new species of ant.

Much of Wallace's interest was taken by the way in which different locations seemed to contain minor species variants, each suited to the location in which it was found. In 1855 Wallace sent back to London a paper titled *On the Law Which Has Regulated the Introduction of New Species*. The paper was clearly influenced by Lyell's view that if geological processes had not differed through history, then immensely slow and long-term processes were at work in nature. Wallace related this idea to his observations on species change: "It would be most unphilosophical to conclude without the strongest evidence that the organic world so intimately connected with it [the inorganic world of Lyell's work] had been subject to other laws which have now ceased to act, and that the extinction and production of species and genera had at some later period suddenly ceased."

Wallace believed that a species could be divided into two or more
variants, and if at some time the original species died out it left its
variant behind as new species. Wallace had developed a theory of
evolution. When Darwin[74] read Wallace's paper he wrote to him: "I 74 4 27
agree to the truth of almost every word of your paper; and I daresay
that you will agree with me that it is very rare to find oneself agreeing
pretty closely with any theoretical paper . . . I can plainly see that we
have thought much alike and to a certain extent have come to similar conclusions."

Three years later Darwin (and nobody else) received Wallace's next paper, *On the Tendency for Varieties to Depart Indefinitely from the*

Original Type. Darwin said it hit him like "a bolt from the blue." Everything he had been thinking about for twenty years was in Wallace's paper. In great haste he put pen to paper to produce a work uncannily like Wallace's. On July 1, 1858, the Linnaean Society in London heard both men's papers. Wallace conceded Darwin's primacy.

In one important way Wallace differed from Darwin. For him evolution could not account for the clearly special nature of human consciousness: "Neither natural selection nor the more general theory of evolution can give any account whatever of the origin of sensational or conscious life. They may teach us how, by chemical, electrical, or higher natural laws, the organised body can be built up, can grow, can reproduce its like; but those laws and that growth cannot even be conceived as endowing the newly-arranged atoms with consciousness." Wallace, like many other eminent Victorian scientists, turned for the answer to spiritualism, becoming a leading apologist for the movement in England. Wallace accepted as valid all manifestations of the psychic world, including ghosts, haunted houses, communication with the dead, levitation, and ectoplasm. In 1882 he joined the new Society for Psychical Research, but refused the post of president.

One of his colleagues at the society was a Liverpool University professor of physics named Oliver Lodge. Like many other scientists, Lodge attempted to use scientific method to provide a philosophical or religious meaning to life and in this way mitigate the apparently godless aspects of science. In 1883 Lodge was asked to examine the case of two department-store salesgirls who appeared capable of thought transference. So as to avoid ambiguity in the test results, Lodge developed the now-well-known "telepathy" cards carrying a circle or triangle or square.

Lodge was also interested in other forms of transference. In 1889, after a number of experiments with lightning conductors and the electrical waves created by sparks, he became involved in developing a sensitive receiver for the very weak signals given out by spark transmitters. In 1889 he said he "came across a curious effect . . . whereby a couple of little knobs in ordinary light contact, not sufficient to transmit a current, became cohered or united at their junction whenever even a minute spark passed, and thus enabled the passage of a current from a weak E.M.F. [voltage] through a galvanometer, until they were broken asunder again, which a light tap sufficed to do."

Lodge noticed that the same phenomenon happened when currents passed through iron filings. Even a relatively weak signal was enough to cause the filings to stick together and pass on the signal. Lodge built a small glass tube filled with the metal particles and later added a "tapper" that would give a series of repeated blows to the glass tube when the filings had cohered, breaking them apart ready to receive the next signal. In 1891 the detector became known as the Branly-Lodge "coherer" (Edouard Branly was a French physicist who turned out to be working on the same principle, in parallel with Lodge).

It was a Canadian engineer who would take things to the next stage. The coherer had severe limitations in that it reacted only to the kinds of signal bursts given off by spark transmitters (the kind Marconi used with his new radio telegraph to send Morse Code dots and dashes). The Canadian, Reginald Fessenden, realized that to send anything more than dots and dashes required an entirely different
form of radio transmission. His experience working for Edison[75] on **75** 23 *48*
dynamos, and then as chief electrician at the Westinghouse[76] Electri- **76** 25 *49*
cal Company was to prove useful. As part of Fessenden's involvement with power generation and electric motors he had worked with alternating current. This is a current which grows to a peak, then decreases, reverses its direction, reaches a peak, then returns to its original state. It completes this cycle thousands of times a second. Fessenden realized that he could use this process to set up continuous wave transmission at a set frequency (so a receiver could be tuned to it), and then use the wave as a carrier of smaller energy inputs from a microphone reacting to sound inputs. The signals from the microphone would modulate the carrier wave. At the receiving end, these modulations could be fed to a loudspeaker diaphragm that would vibrate in sympathy with the fluctuating input and repeat the original sound.

On Christmas Eve 1906 at Brant Rock, Massachusetts, Fessenden described what happened when he put the theory into practice: "The program . . . was as follows: first a short speech by me saying what we were going to do, then some phonograph music, the music on the phonograph being Handel's 'Largo.' . . . finally we wound up by wishing them a Merry Christmas and then saying that we proposed to broadcast again on New Year's Eve." Those listening to this first-ever radio program were asked to write to Fessenden at Brant Rock. Several did so, including the radio-telegraph officers on board ships in the Caribbean. The ships belonged to a company whose fortunes were to be radically changed by Fessenden's work. They were the banana boats of the United Fruit Company.

Bananas had been a highly profitable commodity ever since 1870, when a Bostonian schooner master had returned from taking a party of gold prospectors to Venezuela. On his way home he stopped in Jamaica for repairs and bought 160 bunches of bananas for twenty-five cents a bunch. Later he sold them in Boston for more than ten times as much. By the time of Fessenden's Brant Rock transmission the banana business was flourishing. American banana growers owned nearly a quarter of a million acres in Costa Rica, Cuba, Honduras, Jamaica, Santo Domingo and Colombia. Many of them had considerable influence with these local governments, and the countries were becoming so dependent on the export of the fruit they were beginning to be known as "banana republics."

The reason United Fruit leaped at Fessenden's invention was simple. Bananas are so profitable because while an acre of wheat will yield thirteen hundred pounds of crop, and corn about twenty-eight hundred pounds, an acre of bananas will provide eighteen thousand pounds. Moreover, bananas can be harvested all year round and they grow extremely fast. The problem for United Fruit was that the fruit matured and rotted with equal speed, so the crop had to be dispatched as quickly as possible. Tens of thousands of banana bunches were loaded on trains in plantations and delivered to the docks with the requirement that they be loaded on board ship and dispatched within a few hours of the scheduled time. It took dozens of trainloads to fill a ship's hold, so the scale of the investment at risk was considerable. In 1900 United Fruit (an association of twelve banana firms) owned 11 steamships, 112 miles of railroad, 17 locomotives, 12,000 cattle and 2,000 horses and mules and employed 15,000 people clearing 8,000 acres of jungle a year ready for planting.

In 1908 the value of refrigerated ships[77] was already proven, and 77 56 *116*
United Fruit commissioned the construction of seventeen five-thousand-tonners. The enterprise became increasingly complex as the company began building piers, laying hundreds of miles of railroad tracks, developing dock facilities in a dozen tropical ports and, as the market continued inexorably to grow, organizing the clearance of thousands of acres of jungles for ever-more plantations. The logistical task of organizing this far-flung and fragmented commercial empire with its critical reliance on efficient and timely scheduling was made possible almost at a stroke by Fessenden's radio.

During the early years of the banana trade the world authority on the fruit was a book written in 1882 by a reclusive Swiss botanist (from a family of reclusive botanists) named Alphonse Pyramide de Candolle. His book was called *The Origin of Cultivated Plants* and contained the first detailed botanical description of the banana. Candolle also wrote books on the culture of fruit trees, the age of trees and the dormancy of plants. As a member of Geneva's Grand Council Candolle also introduced the first postage stamps to the canton. In 1843 the Geneva Council voted to introduce (on the English model of only four years earlier) a single-value stamp that would prepay all local postage. Two stamps would be required for postage to another canton. In 1852 this system was extended to the whole of Switzerland.

By the 1860s the international postal situation[78] was close to 78 49 97
chaotic. Countries applied their own rates (some as many as six) to different kinds of mail and different distances of delivery. Errors and losses were endemic. Various attempts were made to arrive at some kind of international postal consensus modeled on the highly successful Austro-Prussian Postal Union, which had been operating since 1850. Eventually, in 1874, Heinrich von Stephan, director of posts for the North German Confederation, managed to organize an

International Postal Congress in Bern, Switzerland. Twenty-one countries sent delegates with the aim of "transforming the entire world into a single postal territory for the reciprocal exchange of the mails." At Bern it was agreed that where possible mail should be prepaid by means of postage stamps and that countries of origin should keep the income from the sale of stamps on the grounds of simple reciprocity: a letter tended to elicit a reply.

One of the other congress decisions settled the cost of different classes of mail. These were to be printed materials, letters and (a new category) postcards. Postcards had first been mooted by von Stephan, who realized that letter-writing was a long-winded affair and that people might want a briefer form of correspondence. He suggested a card carrying a preprinted postage stamp, which would have no need for an envelope. The card itself would be free. This last proved a sticking point and the idea was dropped. It was revived again in 1860 in Austria with the printing of official "correspondence-cards," which proved immensely popular. Half a million were sold in the first month and several million were posted in the first year. Germany followed the example, then Britain. In 1870 cards began to carry printed Christmas greetings. In 1872 Britain authorized private printers to issue cards. In 1889, at the Paris Exhibition, the picture postcard
79 128 *244* burst on the scene. It could be mailed at the top of the Eiffel Tower.[79]
One side of the card carried a lithograph of the tower, the other was left blank for address and greeting. The idea was a huge success and inspired others to follow.

By the end of the nineteenth century the first artists' cards appeared in France, principally displaying the work of poster painters like Boutet and Mucha. In 1900 illustrated cards started appearing in Britain. Here, however, the illustrations took the form of humorous cartoons and caricatures featuring the "day-tripper" at the seaside. Over the next decade the pictures became more sophisticated, particularly with the work of such artists as Phil May. May began at the age of fourteen in the Leeds Grand Theater, helping to paint backcloths. After he had painted several of the actors his talents were noticed and he began to draw for London magazines.

Fig. 17: *An early British postcard showing the usual "naughty" scene of industrial workers on holiday at the beach.*

Eventually May was hired as a cartoonist for the satirical magazine *Punch,* where his work lampooned the politically great and good. *Punch* had begun publication in 1841 at a time when Britain was suffering from the worst excesses of overrapid industrial development. The high hopes of the great Reform Bill of the previous decade had come to nought. The towns were packed with factory workers living in appalling conditions. Corruption among government officials was rife. MPs were venal and self-seeking. The disparities between rich and poor were great and widening. *Punch* entered the fray on behalf of the poor and dispossessed, mercilessly attacking those in power.

In 1843 came an opportunity too good to be missed. Queen Victoria's husband, Prince Albert (not known for his artistic abilities), joined a committee to judge a design competition for frescoes to be painted in the newly built Houses of Parliament. The committee required competition to be inspired by Classical themes or subjects from English history. The works submitted were so bad that *Punch* decided to run its own competition and present its own winning designs. These included cartoons excoriating everybody from factory owners to aristocrats. Meanwhile the real competition triggered a storm of criticism when the winning designs were painted. By 1895 all but one had peeled off the walls or had been covered up. Today none is visible.

The historical subjects of the frescoes suited the style of the Parliament buildings. Ever since 1733 there had been demands for a new seat of government. At one point Buckingham Palace was offered by
80 14 37 William IV. By the end of the eighteenth century the Neoclassical[80] style of buildings like Buckingham Palace had been replaced by Neo-Gothic, which appealed more to the nationalist fervor of a country at war with France. Gothic architecture was held to be of English origin and favored by those who looked back to the golden age of Saxon liberty when the rights of free Englishmen had first been proclaimed. In 1801 the British and Irish legislatures were combined, then in the 1830s (after the Reform Bill) the number of MPs grew to over six hundred, and conditions in the old Parliament became impossible. The situation was exacerbated by the Great Fire of 1834, which destroyed the building.

It was decided that the new Houses would be Neo-Gothic. Their decoration was given to the greatest exponent of the English Gothic Revivial, August Pugin. Pugin's first book stated his position unequivocally. It was titled *The True Principles of Pointed or Christian Architecture,* and it described the direct link between spirituality and design, citing the religious faith of medieval builders as the prime example. Gothic architecture for Pugin was faith writ large in stone. Pugin designed everything in Parliament, from gargoyles to ceiling moldings, woodwork, carpets, metalwork, furniture, carvings, glass and everything in the great ceremonial chambers such as the House of Lords, perhaps the greatest pseudomedieval interior ever built. Pugin's theme in the Lords stressed the medieval origins of the parliamentary system. He included bronze statues of the barons who forced King John to sign Magna Carta, angels modeled on those of the 1388 Westminster Hall, and a three-part, raised medieval throne. The whole extraordinary edifice was described by American author Nathaniel Hawthorne as "gravely gorgeous."

Ironically, given the nationalist fervor behind the British decision, the Gothic Revival had begun in Germany with the recent Romantic movement, whose aesthetic leader was an ex-medical student and writer named J. G. Herder.[81] Herder first taught at the Cathedral **81** 93 *174*
school in Riga, Estonia, in 1764, where he began to write about German literature. Five years later he traveled in France and Holland. In 1770, visiting Strasbourg for an eye operation, Herder met another young student who was there to complete his legal education. Johann Wolfgang von Goethe[82] was to prove a seminal influence on **82** 22 *47*
Herder, persuading him of his position as an important German au- **82** 69 *141*
thor. Herder in turn stimulated Goethe to give up the law and concentrate on literature.

Herder's interest in folk-poetry and ancient languages led him to a deepening involvement with the German past and a growing understanding of the particular nature of German culture. From the German art historian Johann Winckelmann Herder borrowed the idea of the need to understand the historical context of cultural expression. Herder became known as the founder of the "historical outlook," the originator of the concept of humankind united at all times and in all conditions, and the idea that man existed as part of nature. These were the first guiding principles of what would become known as Romanticism. One of Herder's essays dealt with the development of language. In *Treatise on the Origin of Language,* published in 1770, he traced the development of language through history, describing language as a divine gift capable of expressing the most fundamental awareness of God's revelation. The value of ancient languages was that because they were archaic they were the purest, least alloyed by the effects of historical development. Herder argued that the oldest examples of language would reveal the most meaningful understanding of humanity's origins.

Not surprisingly, when a major third-century epic poem in Gaelic was discovered Herder and the other Romantics received it like revealed truth. The poem was called "Fingal, an Ancient Epic Poem in Six Books, Together with Several Other Poems Composed by Ossian the Son of Fingal." In it Herder and the other Romantics found the answer to their prayers. Here was a vivid portrait of third-century

Celtic society that revealed a culture as great as that of Rome or Greece. The poem served as a beacon to those who sought an identity and a tradition for German culture. "Ossian" was the lyrical expression of a simple peasant, living in a society before the emergence of class, wealth or any of the other destructive artificial aspects of modern Enlightenment (French) civilization. For Herder the poem was a clarion call from the noble savage to all Germans, and in his essay "Ossian and Ancient Folk-Poetry" he rallied every Romantic to the cause.

Most unfortunately the epic "Ossian" was an epic fake put together by a young, literary Gaelic-speaking Scottish Highlander named James Macpherson, who was concerned at the fact that Gaelic culture might be dying and was anxious to preserve it. Macpherson toured the Scottish Highlands and Islands collecting ancient Gaelic tales, songs and poems, and in 1761 he published them in the "Ossian" collection. Macpherson "creatively restored" the fragments he had written down, adding his own material and assembling them into an epic structure, biblical in diction and classical in style, creating a Celtic twilight zone that had never really existed, full of the supernatural, the mysterious, the heroic. A world where man was at one with nature. A world perfectly designed for the Romantic mind.

The reason for the forgery lay in the changing social conditions of Scotland early in the eighteenth century. Not long after the Union with England in 1707, resentment still smoldered north of the border, where the Scottish government had been dissolved and Scots felt themselves under-represented in London. In 1715 a Scottish rising against the English was headed by James Stuart, who claimed the English throne and united the clans against the English occupying forces. After the rebellion was savagely put down the English took pains to make sure it would never happen again. Gaelic was banned in schools. To facilitate the movement of English garrison troops military roads were built through the Scottish Highlands by Marshal Wade (in whose honor an extra verse was added to the national anthem). The Society in Scotland for Propagating Christian Knowledge did all it could to undermine the Highland Catholic faith in Scotland and replace it with Protestantism, especially in the Lowland areas where the new Union was promoting economic growth and turning places like Glasgow into commercial entrepôts. The added effect of the industrialization of the Lowlands was to split Scotland in two and weaken further the position of the economically backward Highlands.

The Stuarts, exiled on the Continent, continued to claim the English throne. James moved to Rome, living on handouts from the Vatican and the Italian nobility. In 1719 he married the daughter of the Polish royal house and their Roman residence in Palazzo Muti became a center for Jacobite disaffection. In 1720 James's wife gave birth to a son christened Charles Edward Louis John Casimir Silvester Maria Stuart. Wrapped in the robes of the Prince of Wales and laid under a royal canopy of state, the infant was visited by hundreds of Jacobite well-wishers and treated in every way as if he really were the heir to the English throne. For the next twenty-five years Charles

Fig. 18: *The young Chevalier, Bonnie Prince Charlie, wearing the Royal Order of the Garter (to which he was not entitled).*

grew up with the myth of kingship, charming and good-looking, a natural athlete, speaking English with a foreign accent.

In 1740, with the British at war with France, the Jacobites decided (with French support) to try again. James was too old to lead so Charles was to be sent in his stead, backed by a French fleet and troopships. Bad weather scattered the fleet and Louis's support for Charles's adventure melted away. In an act of supreme folly Charles elected to go on alone, landing in the Scottish Western Isles in July 1745. Hailed by the Highlanders as their savior, "Bonnie Prince Charlie" miraculously got his rag-tag Highland army as far south as Derby, fifty miles from London, before they were turned back and chased north by the British army. The decisive battle was fought at Culloden on April 16, 1746, when one thousand badly equipped and ill-fed Highlanders wielding claymores were massacred by nine thousand disciplined British redcoats using artillery. Charlie fled back to the Islands and was soon on the run.

The systematic destruction of the Highlanders began in earnest. British soldiers went on a rampage, looting, raping, killing and burning property. Charlie escaped, returned to the Continent, and for the rest of his life moved from borrowed palace to palace, existing on charity and becoming a chronic alcoholic. Toward the end he would play Scottish airs on his cello and drink himself insensible every night, weeping at his lost greatness, eventually dying in Rome in 1788 at the age of sixty-eight.

Meanwhile in Scotland Highlanders were forbidden to carry arms, play the bagpipes or wear tartan. The estates of the fourteen most prominent Highland clan chieftains were seized by the English crown. All ancient feudal jurisdictions were abolished, and the hereditary authority of the chiefs was destroyed forever. Thirteen years later the Highlands were considered to have been subdued. A new, sentimental view of the ancient Highland world began to find expression among the English chattering classes. The king visited Scotland wearing a kilt.

Highlanders deserted Scotland in thousands. In 1775, one of the émigrés was a fifty-three-year-old woman named Flora Macdonald. As a young girl of twenty-three she had played a significant role in the 1745 rebellion, smuggling Charlie (disguised as a maid named "Betty Burke") out from under the noses of the British troops and rowing him to safety on the island of Skye. Such was her fame for this act that many years later Samuel Johnson was to say: "Her name will be mentioned in history, and if courage and fidelity be virtues, mentioned with honor."

When Flora and her husband left Scotland[83] they followed in the **83** 98 *181*
footsteps of their fellow-countrymen and took ship for colonial America, the land of plenty and above all of freedom. Although British ill-treatment had been instrumental in the Highlanders' decision to leave, it is more likely that poverty was the principal factor. With the destruction of the old feudal system, in which chieftains had taken their rents in kind or in service, the new conditions encouraged landlords to charge money for rent. Many Highlanders could not pay. In the winter of 1771 and the disastrously cold and wet spring that followed much of their livestock died. Population pressure played a part, as Highland women were noted for their ability to bear as many as twenty children. The introduction of sheep also brought large-scale evictions. Wherever they went on their tour of Scotland in 1773 Boswell and Johnson found people preparing to leave. Farewell laments were being sung in every village.

By far the largest number of Highlanders went (as did Flora) to North Carolina. The land grant of fifty acres for every immigrant to the colony was an attractive prospect for people who left Scotland with virtually nothing. In spite of their past experiences the Highlanders remained Loyalist in the impending conflict between America and Britain.

For the British, North Carolina was of particular importance because it provided much of Britain's naval stores. At the beginning of the eighteenth century, when Britain had been at war, the only source of naval stores had been the monopolistic Swedish Tar Company, and it was to avoid a repetition of this situation, and as part of the general mercantilist thrust to achieve economic independence, that Britain had passed the 1705 Naval Stores Bounty Act, authorizing subsidies for all domestic production of naval stores.

"Naval stores" was a term that described materials used in various ways to waterproof ships. Stores included tar, pitch, rosin and turpentine. Tar was used to preserve ropes, hulls were caulked with pitch, and turpentine was used to thin paint and to protect woodwork (although it was more often used as a medicine, administered orally for tapeworm, rubbed on for rheumatism and bronchitis, put on wounds as an antiseptic, or ingested as a purgative). In an era of wooden ships, naval stores were essential to maintaining naval capability, so when the first North Carolina colonists discovered the extensive areas of long-leaf pine trees on the colony's coastal plain, the area rapidly became Britain's new source. Turpentine was distilled from the rosin that flowed from a cut pine tree; tar was produced by cooking pieces of pinewood in kilns; pitch was made by boiling the tar in caldrons or open pits. By the second half of the eighteenth century 70 percent of the tar, 50 percent of the turpentine and 20 percent of the pitch imported to Britain came from North Carolina, earning for the colonists the nickname "tarheels."

When the American War of Independence ended new sources of British turpentine had to be found, not least because of the new craze for Chinese lacquered furniture. Fine lacquer work reached its zenith in 1680 at the Imperial Palace in Peking and by mid-eighteenth century was being imported to Europe by the Dutch. Lacquer was astronomically expensive and came in the form of tea tables, panels, chests, screens, snuffboxes, trinkets, even coaches. "Japanning," as the technique came to be known in England, involved covering the objects in several layers of clear varnish and then painting or gilding them and adding further coats of varnish to produce a lustrous effect. At some time around 1730 a Welshman named Thomas Allgood developed a cheap substitute. His ingredients were linseed oil, umber (brown oxide of iron) and litharge (a lead monoxide). The mixture was heated and diluted with turpentine obtained from oil shale in the hills around Pontypool where Allgood lived.

Allgood applied his new "Pontypool Japan" to metal. Wood was scarce and the use of timber was severely restricted to shipbuilding, but Pontypool was also the location of one of the best ironworks in Europe, run by John Hanbury and employing the Algood family. The works produced large quantities of rolled sheets of iron and used the latest hot-rolling processes. In order to prevent rusting iron sheets were "tinned" by being dipped in molten tin. Thomas Allgood's grandfather (also named Thomas) had learned his tinning during a visit to Saxony and Bohemia, the European center of tin-making at the time. Thomas Junior applied several layers of his "Pontypool Japan" to tin plates and stoved them for hours in an oven. The finished product was then shaped into trays, tobacco boxes, candlesticks, coffeepots, tea caddies, sugar canisters, kettles, pans, basins, boxes and other household articles. The Royal Navy purchased some for use as bread containers.

A similarly naval use for tinplate was planned in France where in
1661 the new first minister, Jean-Baptiste Colbert,[84] had recently 84 138 *255*
taken over an economy that was in ruins. In an attempt to set French industry on its feet Colbert began by importing foreign craftsmen to teach Frenchmen how to set up their own production units. Colbert brought cloth-makers from Holland, tar-makers from Sweden, lace- and glass-makers from Italy, goldsmiths from England, leatherworkers from Russia, sugar-refiners from Germany and hatmakers and weavers from Spain. However, attempts to persuade tin-makers from Saxony to emigrate to France under very favorable conditions (tax-free subsidies and automatic rights of French citizenship) failed.

One of Colbert's other (and more successful) economic reforms involved the navy. When Colbert took over, the French navy consisted of only eighteen warships, some over twenty years old. Six thousand French sailors were serving in foreign navies. Not a single mast could be found in France's arsenals and storehouses. Ten years later, thanks to Colbert, France possessed 190 vessels, of which 120 were fully equipped warships. Colbert bought masts in Savoy, tar in Prussia, wood in Poland, naval stores and munitions in Holland. He imported Dutch carpenters and English naval architects, built or refurbished shipyards in Brest, Toulon and Rochefort, opened schools of hydrography at Rochefort and Dieppe, and established officer training courses at Rochefort, St. Malo, Toulon and Brest. In addition he reformed conditions for enlisted men by requiring a man to serve only one year in three, and provided free education for children, homes for wounded or disabled sailors and family allowances.

In 1667 the Toulon Arsenal employed a painter named Pierre Puget as artistic director, whose prime responsibility was ship decoration. Puget's style did not appeal to Colbert, one of whose administrators said he felt they were "more relevant to the decoration of a palace than to ships." Puget specialized in carved and gilded sternpieces, which including a remarkable variety of caryatids, atalantes and triton figures. These showed the powerful influence of the man who had been Puget's mentor (and employer) in the early years of his artistic development, the Italian master Pietro da Cortona, with whom Puget had worked in Rome and Florence.

By the time Cortona was employed by the dukes of Tuscany, in 1637, he was already successful and well-known as a painter and architect, having worked for the Sachetti and Barberini, two of the most powerful families in Rome. It was while passing through Florence with Cardinal Sachetti that Cortona was persuaded by Duke
85 110 *208* Ferdinando II[85] of Tuscany to join the many Florentine artists employed in embellishing the newly refurbished ducal residence (the Pitti Palace). Ten years later Cortona was again employed to decorate the ceilings of the ducal apartments with scenes from Greek heavenly mythology, as a result of which the apartments became known as the Planetary Rooms. On one of the ceilings Cortona artfully included references to the "Medicean" satellites, the moons of Jupiter discov-
86 105 *200* ered by Ferdinando's mentor Galileo[86] and named after the ducal family in recognition of its support of science. Ferdinando and his brother Leopoldo were both admirers of Galileo and had given him a formal burial when the church had forbidden him any monument. The two brothers also established a science academy, carried out their own experiments and closely followed the work of Evangelista Torricelli, Galileo's pupil, during his work on the vacuum in 1643.

The news of Torricelli's discovery took Europe by storm. Apart from anything else it stirred up a major theological row. If the vacuum was an absence of everything was God also absent from it? Scientists concentrated on more mundane aspects. In 1661 Robert
87 53 *114* Boyle[87] published Boyle's Law after he had used a vacuum pump to
87 65 *135* show that at a constant temperature the volume of a gas is inversely

proportional to its pressure. In France this became known as Mariotte's Law, after the name of the French scientist who relied heavily on Boyle's work when in 1679 he produced his own: *On the Nature of Air.* Edmé Mariotte indulged in semiplagiarism more than once, eliciting complaints from several scientists, including Christiaan Huygens,[88] the inventor of the pendulum clock. Mariotte had a similar, **88** 52 *114*
"symbiotic" relationship with Pierre Perrault, an ex–tax collector in Paris who had been caught with his hand in the till and who in 1674 published a book titled *The Origin of Springs.* Perrault had measured the drainage characteristics and annual rainfall in the Seine Basin, examined runoff processes in plants, and concluded that only one-sixth of the annual rainfall was necessary to sustain the Seine's flow. This was the first experiment that showed that rivers owed their flow to rainfall. Mariotte extended Perrault's findings by setting up a series of weather-reporting stations across France and published his own book, *The Movement of Waters,* in 1686.

Pierre Perrault came from a talented family. In 1667 his younger brother Claude designed the colonnade in the Louvre. His older brother Charles was a member of the new French Academy of Sciences and at the age of thirty-five became Jean-Baptiste Colbert's factotum, minister for culture, member of the French Academy, and adviser to poets, including Racine. He also had the unenviable task of keeping the accounts during the construction of Versailles.[89] **89** 137 *254*

Charles Perrault goes down in history principally for having written a collection of children's stories in 1697 under the title *Tales Told by Mother Goose,* which included "Little Red Riding Hood," "Puss in Boots," "Cinderella," "Tom Thumb" and "Sleeping Beauty." The collection was later translated into English and published as *Mother Goose's Melody.* In the same year Perrault also triggered a great literary debate with the publication of a poem in which he denigrated the value of classical literature, claiming that modern writers were better and describing Plato as "boring." A furious argument broke out within the French literary establishment and then spread across the Channel to England. There, a retired statesman, Sir William Temple, rose to Perrault's defense. This was a time when recent mathematical and scientific discoveries, economic advance and exploration were all contributing to a general rise in confidence that made such people as Temple feel that the modern age had much of value to offer. Temple reckoned without opposition from the entrenched establishment of the church and from the old guard at Oxford and Cambridge.

Fortunately, he had in his employ a young man who was to prove equal to the task of defending Temple's position. His name was
90 124 *232* Jonathan Swift[90] and in 1704, in "The Tale of a Tub," Swift employed his devastating satirical wit in an attack on pedantry and religion, referring to the clergy: "Who that sees a little paltry mortal, droning and dreaming and drivelling to a multitude, can think it agreeable to common good sense that either Heaven or Hell should be put to the trouble of influence or inspection upon what he is about?" As for officials: "If one of them be trimmed up with a gold chain and a red gown and a white rod and a great horse, it is called Lord Mayor. If certain ermines and furs be placed in a certain position we style them a judge; and so, an apt conjunction of lawn and black satin we entitle a bishop."

Not surprisingly this approach to authority lost Swift all chance of a well-paid sinecure within the church. Temple died leaving him only one hundred pounds per annum, and Swift was in dire straits until the Lord Chief Justice in Dublin, George Berkeley, took him on as chaplain. Berkeley had spent twenty-four years in Dublin at Trinity College, studying Classics, Hebrew, logic and theology. He had also traveled Europe, become dean of Derry, and married the daughter of the speaker of the Irish House of Commons. In 1728 he had gone to America, where his scheme was to establish a university in Bermuda to educate young men working in the plantations and train native Americans as missionaries. When the money for the foundation was not forthcoming Berkeley spent five years in New England, helping to found the Philosophical Society and leaving his mark on American education (several cities were named after him, including the university city in California). After leaving his books and his estate to a recently founded college in New Haven (it would become Yale), he returned to Ireland as bishop of Cloyne.

In 1709 at an early age Berkeley had already published *A New Theory of Vision,* in which he developed the idea, later to be incorporated into his associationist philosophy, that light and color were merely tactile experiences to be interpreted and given meaning by the brain. This act of interpretation relied on the association of ideas, which was in turn learned through experience.

Berkeley's observations on vision were taken up at the end of the
eighteenth century by Thomas Young,[91] a precocious talent who was **91** 140 *259*
said to have read the Bible twice by the age of four. Before he was twenty Young had learned French, Italian, Latin, Greek, Hebrew, Syriac, Chaldee, Samaritan, Arabic, Persian, Turkish and Ethiopian, as well as entomology, botany and philosophy. In 1793 the twenty-year-old Young entered St. Bartholomew's Hospital in London as a medical student. That same year he wrote his first important scientific paper, *Observations on Vision.* In it he stated the first modern optical theory of color vision, noting that the retina reacted to colors in terms only of variable amounts of the three primary colors: red, green and violet. By 1801 Young was professor of natural philosophy at the Royal Institution, where he lectured on acoustics, optics, gravitation, astronomy, tide, capillary attraction, electricity, hydrodynamics, measurement and other things.

In 1814 this polymath turned his attention to hieroglyphics when a friend brought back fragments of papyrus from Egypt. He then began work on the Rosetta Stone, a monument carrying inscriptions in Greek, and two forms of Egyptian (demotic and hieroglyphic). By comparing the Greek and demotic Young was able to locate the names "Alexander" and "Alexandria." He also noticed the frequent repetition of a sign that turned out to be "and." Reasoning that Egyptian scribes would use the phonetic form of foreign names and that proper names would be surrounded by a ring (forming a "cartouche"), he identified the names of Ptolemy and Cleopatra. Young's work prepared the ground for the full decipherment of hieroglyphics by Jacques-Joseph Champollion less than a decade later.

The Rosetta Stone, on which both men worked, was a slab of polished black basalt three feet nine inches long and two feet four

inches wide discovered in 1799 by French soldiers repairing the ruined Fort Rashid, close to the town of Rosetta in the Nile Delta, during the Napoleonic occupation of Egypt. After the British had driven Napoleon out of Egypt, the Stone was removed and taken back to London. Other hieroglyphic source materials (including inscriptions on temples, obelisks and stele) had also become available between 1809 and 1816 with the publication by the French authorities of *The*
92 45 *87* *Description of Egypt,*[92] a giant book full of drawings, measurements and reports on the country prepared at Napoleon's command during the occupation.

The work of publication was superintended by Nicolas-Jacques Conté, who spent three years in Egypt as a kind of quartermaster, keeping the military and scientific staff supplied with everything from swords to magnifying glasses and surgical instruments. Before Conté's departure for Egypt he had been head of the new Paris Conservatory of Arts and Crafts and had made his reputation by solving the great French pencil problem. The Napoleonic Wars and the trade blockade that accompanied them had curtailed French imports, including that of pencils. Conté developed a method for refining graphite by mixing it with clay and making it smooth enough to use as pencil lead. The pencils he produced still bear Conté's name today.

At this time Conté was also known for his work at the Aerostatic Institute at Meudon, near Paris, where he helped found the new Aerostatic Corps of the Artillery Service. The Corps mission was to fly in reconnoitering balloons and observe enemy troop movements. For some reason Napoleon took against the idea, and in 1802 on his return from Egypt disbanded the force. Meanwhile others had been excited by the possibilities of flight. Benjamin Franklin had seen the original Montgolfier balloon experiments in 1783 and agitated in America for similar work to be undertaken there.

However, it was not until the beginning of the American Civil War that American aeronauts would take to the sky. At the time, they were referred to as "professors," the most famous of whom was Thaddeus Lowe. In 1859 he had built the largest-ever balloon (named *Enterprise*, it was 200 feet high and 130 feet in diameter)

Fig. 19: *1861. A Union balloon prepares to ascend. In the absence of a telegraph wire, note the white flag for signaling to the ground.*

with the intention of crossing the Atlantic. Unfortunately, the wind blew him the wrong way and he landed in South Carolina, where he was arrested as a Yankee spy. By 1862 Lowe was a real spy, working for General McClellan's Army of the Potomac, using a number of his balloons to rise to five thousand feet and report via telegraph wire to the ground the disposition and activity of Confederate forces. Lowe's most famous and successful effort was at the battle of Chickahominy, on June 1, 1862, when the *Times* of London reported that he had hovered two thousand feet above the battlefield during the whole of the engagement.

General George McClellan was well aware of the value of spying and recruited a man who had worked for him during McClellan's prewar time as president of the Illinois Central Railroad (whose legal counsel was Abraham Lincoln). Earlier, in 1849, McClellan's new employee had become Chicago's first full-time detective. In 1850 he had set up his own detective agency, whose motto was "We Never

Sleep." Allan Pinkerton's firm was successful because he collected dossiers on known criminals and was probably the first detective to develop the idea of criminal MO (modus operandi). Pinkerton was also a master of disguise, with a large collection of wigs and costumes. As soon as McClellan was appointed to his military post he brought Pinkerton to Washington, where Lincoln asked him to organize a secret service department that would gather information on the social and political activities of suspect individuals in the city.

After the Civil War ended, one of Pinkerton's greatest successes (apart from catching Butch Cassidy and the Sundance Kid) involved a group of Irish-American anarchists named the Molly Maguires. The group carried out assassinations and bombings in the Pennsylvania coalfields, where relations between mine owners and miners were at an all-time low. In 1873 the president of the Philadelphia and Reading Railroad, which had lucrative contracts to haul coal, hired Pinkerton and asked him to infiltrate the Mollies. Pinkerton chose a recent Irish immigrant named James McParlan. Over the next two years McParlan managed to gain the trust of the Mollies, to the extent that they began to ask him to carry out assassinations. In order to avoid this, McParlan pretended to have a drinking problem while continuing to write secret letters to Pinkerton reporting on the Mollies' activities. In 1875, fearing that his cover was blown, McParlan managed to escape by the skin of his teeth and Pinkerton sent him to Denver as head of the local agency there (as well as to recover his health). Meanwhile, partly because of McParlan's work, the Mollies were rounded up and thirteen of them hanged.

In 1913 William Burns (at that time America's greatest detective, with his own agency) visited London and recounted the tale of McParlan's adventures to a fellow crime enthusiast whose own detective methods were to become so well-known that in 1924 the *Illustrated London News* referred to him as the man who had "evolved and disseminated successfully the constructive method in use today in all Criminal Investigation Departments. Poisons, hand-writing, stains, dust, footprints, traces of wheels, the shape and position of wounds, and therefore the probable shape of the weapon which caused them; the theory of cryptograms, all these and many other excellent methods . . . are now part and parcel of every detective's scientific equipment."

In 1914 the English crime enthusiast wrote about McParlan in a book he titled *The Valley of Fear.* It was the last of Conan Doyle's novels in which Sherlock Holmes would say "Elementary, my dear Watson!"

CHAPTER 7

A Special Place

In 1984 the job of every detective was changed when a British scientist named Alex Jeffreys began the investigation of a group of genes responsible for the production of a protein that carries oxygen into muscle tissue. He was studying DNA as part of his research into the "hypervariable regions" discovered in the United States four years earlier.

The genetic code in these regions of DNA differs markedly between individual human beings, since no two people except identical twins share the same set of hypervariable regions. The regions are short sequences of DNA repeated many times over. DNA itself consists of a long sequence made up of four bases: adenine, guanine, cytosine and thymine (abbreviated to A, G, C and T). The DNA molecule has a double, threadlike structure made up of these bases, which looks like two interlocking spiral chains (a double helix). The helix is held together by the chemical attraction between pairs of bases: adenine is attracted to thymine and cytosine to guanine. If one of the helix threads is, for instance, a sequence of bases, AATTCGTA, the matching thread sequence will be TTAAGCAT.

Long stretches of the DNA chain are the same in all humans, which accounts for the fact that we all share common features, such as two eyes, four limbs and two feet. The differing hypervariable regions appear interspersed in the DNA chain, repeated over and over again. Jeffreys discovered very short "core" base sequences, ten to fifteen bases long, within the hypervariable regions and common to many of them. This invariable segment within a variable sequence was in effect a genetic marker that would flag the presence of a hypervariable region. Jeffreys isolated core sequences and cloned them many times over to produce large quantities. He then labeled the marker sequences with radioactive chemicals. Since a DNA sample could be "denatured" (heated to cause the two strands of its helix to separate), the short genetic-marker sequences could then find their match (A binding with T and C with G) in one of the single strands.

The bonded bases could then be made visible by exposing a cut-up DNA strand and its attached radioactive marker sequences to a film. When the film was processed the radioactive markers showed up as dark bands. Since the markers identified hypervariable regions (different in all humans except identical twins), the dark bands on the exposed film would identify a single individual very much more specifically than a conventional fingerprint. Since a DNA sample can be derived from any human cell (for example from hair, skin, semen or blood), DNA fingerprinting has proved tremendously effective ever since its first use in 1987 in a British criminal case, when it was used to identify a rapist. Since then the technique has had a dramatic impact on the pursuit of justice.

Basic to the process of DNA fingerprinting is the way in which the pieces of DNA to be marked are separated out according to the length of the strands. This involves a technique originally developed by a Swedish scientist named Arne Tiselius, who in 1925 began separating proteins. Tiselius was assisting The Sveborg, who had developed a centrifuge that as it spun caused the lighter or smaller proteins to move out to the edge of the serum in which they were present. Tiselius noticed that proteins sorted out in this way would often be mixed up, making exact identification impossible. To get around this problem Tiselius developed a new process called "electrophoresis," which would become essential to the development of biochemistry. Tiselius put the molecules under investigation in a gel and subjected the gel to an electric charge. The effect of the charge was to repel the molecules. The lighter and smaller the molecules, the further they went. It was this work that gained Tiselius the Nobel

Prize, "For his discovery of the complex nature of molecules occurring in blood serum."

Electrophoresis separates molecules by size or weight into different bands along a glass tube containing the gel. In order to see and photograph these bands, Tiselius used a technique originally invented by a Viennese physicist named August Toepler. It was called "schlieren ['smear'] photography" and it revealed the bands of different protein concentrations because of the changes caused by the bands to the refraction of light passing through the glass tube. In the 1880s Toepler's original use of the schlieren technique had been to show the shock-wave patterns caused by explosions or created by the movement of projectiles.

It was this latter capability of schlieren photography that was to interest a Hungarian mechanical engineer and ex-artillery cadet named Theodor von Karman. Von Karman's father had been knighted by Emperor Franz Joseph for services to education, and Theodor followed in the same academic footsteps, winning a fellowship from the Hungarian Academy of Sciences to visit the German University of Gottingen, where he studied under Ludwig Prandtl. Prandtl was already famous for his work on aerodynamics and for his breakthrough discovery of the boundary layer of air passing over the surface of a wing, the study of which revealed much valuable data about drag and lift. In 1914 Prandtl also discovered the vortices formed when air curls over the tip of a wing and trails behind the aircraft, causing extra drag. Prandtl's mathematics helped wing designers minimize this vortex drag effect.

One major phenomenon was left for Theodor von Karman to explain. On occasion, when airflow broke away from a body, it formed a series of vortices known as a "vortex street." Von Karman showed that these vortices formed alternately at the top and bottom of the body into two vortex trails. If the vortices at top and bottom formed sequentially the effect was stable. If both vortices were generated in parallel the effect was to cause instability and set up cycles of vibration. Von Karman was proved spectacularly right on November 7, 1940, when the new Tacoma Narrows suspension bridge at Puget Sound collapsed after a half-hour of violent oscillations (filmed by a passing member of the University of Washington faculty) generated by a forty-two-mile-an-hour wind. With the use of a model, von Karman showed that the collapse had been caused by a vortex street being shed by the solid sidewall of the bridge. When the oscillations of the bridge were moving in synchrony with the parallel vortices the bridge fell apart. As a result of von Karman's report all suspension bridge sidewalls were subsequently slotted to prevent pressure buildup.

Von Karman's interest in airflow also led him to the study of aerodynamics, and in 1930 he moved to the California Institute of Technology in Pasadena where he helped to set up one of the world's most advanced aerodynamics centers, the Jet Propulsion Laboratory. JPL would carry out much of the leading work on supersonic flight and the development of rockets and spacecraft over the following decades, using schlieren photography to observe the vehicles' behavior.

Earlier in his life von Karman had directed the Austro-Hungarian Air Force research laboratory, where he studied the behavior of propellers and armament. Part of his mandate was to find a method by which pilots could shoot machine guns through their propellers without hitting the revolving blades. At one point early in World War I a Dutch engineer named Anthony Fokker visited von Karman's lab, and the two of them discussed the matter. Fokker knew the answer to the problem, because in April 1915 a Morane-Saulnier monoplane piloted by French pilot Roland Garros had been forced down near Ingelmunster in Germany. On board the aircraft was a machine gun positioned to fire directly ahead and a propeller whose blades were protected by wedges of steel plate to deflect bullets. Fokker immediately devised a way by which the propeller would control the firing of the gun through an interrupter mechanism that fired a bullet only when a propeller blade was not in the way.

When the German Air Force installed Fokker's device it found itself with the world's first fighter planes. All the pilot had to do was point his aircraft at the enemy and pull the trigger. In 1916 when one of the new German fighter planes was being delivered from the factory the pilot inadvertently landed at a British airfield in France. The

British rapidly copied the interrupter mechanism and the way was clear for dogfights. By 1917 aerial combat was commonplace, and the fighter ace was catching the public imagination. In France a pilot became an ace after his fifth kill. Germany required ten. Greatest of all German daredevils was Manfred von Richthofen, the son of a cavalry officer. Manfred was a handsome young Prussian aristocrat whose favorite occupations were hunting and drinking champagne. In 1916 he was assigned to one of the new fighter squadrons and early in 1917 had his own command. Von Richthofen painted his aircraft bright red, earning himself the nickname "Red Baron." By April 1916 von Richthofen had achieved fifty-two kills (his final total would be eighty) and had become a national hero. The German propaganda ministry churned out millions of photographs of him, he received

Fig. 20: *The daredevil Red Baron at the height of his fame in 1918.*

sacks of fan mail and was obliged to make public appearances in factories to denounce communism. By now von Richthofen had a Fokker fighter and all the planes in his squadron had been painted red. Manfred and his squadron were known to their British counterparts as "von Richthofen's Flying Circus." His most famous remark was: "When I have shot down an Englishman my hunting passion is satisfied for a quarter of an hour."

Manfred's great-uncle Ferdinand was professor of geography at the University of Leipzig, and by 1883 he had spent several years as a geologist traveling in Sri Lanka, Japan, Taiwan, the Philippines, Java, California and China (about which he wrote the first definitive geographical description). Ferdinand von Richthofen was the first to bridge the gap between geology and geography. He did so by dividing geography into two fields: special geography (primarily descriptive) and general geography (primarily analytical). Von Richthofen described special geography: "Every area on the earth, no matter how small, whether a continent, a small island, or a naturally bounded inland area, an artificially bounded state, a mountain, a river basin or a sea, is examined as a grouping of smaller unit areas." The description of each of these units in an area became known as "chorography." A synthesis of special and general geography permitted von Richthofen to analyze the interaction between different elements in a geographical area. This also revealed how human presence affected the environment and took into consideration such factors as the distribution of population, races, languages, frontiers, settlements, industries, religions, trade centers, communications routes and products. This kind of study was known as "chorology."

The inclusion in chorology of historical analysis of the effects of human intervention over time was the idea of a German scholar and historian named Carl Ritter, professor of history at Berlin from 1820 until his death in 1859 and founder of the Royal Geographical Society of Berlin. Ritter proposed that the structure of a country was an important element in the development of its inhabitants. He widened the scope of geographical studies: The "science aims at nothing less than to embrace the most complete and the most cosmic view of the Earth to sum up and organise into a beautiful unity all that we know

of the globe . . . and shows the connection of this unified whole with Man and with Man's Creator." The aim for Ritter was to seek the underlying laws uniting the diversity of nature with humankind. In this he was developing the ideas that had galvanized Europe a few decades earlier, at the beginning of the Romantic movement, thanks
93 81 *153* to J. G. Herder.[93]

In *Plastic Art,* published in 1778, Herder sought a psychobiological explanation of aesthetic reactions and introduced the same relativist concepts linking humans and their environment that Ritter and von Richthofen would adopt later. For Herder sense perception was linked to environment. Greenlanders, Herder stated, had no beauty and hence no sense of beauty because their climate was not conducive to the development of beauty. He applied the same argument to history. The arts were a product of their time, defined by the contemporary environment and the physical temperament of the race. One effect of Herder's recognition that each period had unique qualities was the rehabilitation of the Gothic style. Because of this Herder virtually triggered the Gothic Revival that was to sweep Europe in the nineteenth century. For Herder the close links between humans and their location showed that humans were an intrinsic part of Nature itself.

This was the view of Johann Joachim Winckelmann, the founder of the study of art history. In 1764 Winckelmann wrote *History of An-*
94 70 *142* *cient Art,* a work that fundamentally influenced Herder, Schelling,[94]
Goethe, Hegel and many other Romantics. Ten years earlier, after an uneventful life, at the age of thirty-eight Winckelmann had arrived in Rome, settling in the artists' quarter near Piazza di Spagna. In 1758 he entered the service of Cardinal Albani, an avid collector of antiques. Winckelmann devoted himself to a study of these *objets d'art* and in 1762 published *Observations on the Architecture of the Ancients.* This included a description of the recently unearthed city of Pompeii. Since 1748 the gradual excavation of Pompeii and Herculaneum had stupefied Europe. Every day more artifacts and buildings were revealed. Drawing on the new discoveries, in *History of Ancient Art* Winckelmann developed the argument that the ancient world and especially Greece had enjoyed a uniquely "creative" environment.

The Classical spirit had been born of a place whose natural beauty and temperate climate had fostered beauty. For Winckelmann the Greeks represented the ideal, and he painted a glowing picture of them and their artistic genius. Winckelmann's book was a comprehensive analysis of all that was known about Classical art. Winckelmann presented Greece as the origin of modern culture, and put forward the idea, so passionately taken up by the Romantics, that the only way to understand any period in history and its art was to attempt to understand what it had been to live at the time.

One of Winckelmann's friends in Rome was the artist Raphael Mengs. In 1763 Mengs introduced him to an extraordinary young woman painter recently arrived from Switzerland. She was twenty-two-year-old Angelica Kauffmann, and she had just spent time in

Fig. 21: *Winckelmann's inspiration, Pompeii, discovered by workmen digging a well for Charles III of Naples, here inspecting the excavations in 1751.*

Parma, Bologna and Florence, studying and copying the masters. Kauffmann had been a child prodigy, painting the bishop of Como when she was only twelve. She was beautiful, sang well and was an accomplished clavichord player. By the time she arrived in Rome she was feted wherever she went, welcomed by the highest in the land. Winckelmann taught her everything he knew. In return she painted his portrait. It was one of her best.

A year after she arrived Kauffmann visited Naples, where many of the foreign visitors were avid for her to paint them. In 1766 she met the wife of the English resident in Venice, who suggested she go to London. In 1766 she took that city by storm. She had arrived at a brilliant moment in British art. In 1768 the Royal Academy was founded and by 1770 there were so many art exhibitions in London, attracting so many viewers and buyers, that it was difficult to move easily through the streets. Horace Walpole said: "After gaming, the folly of the day is pictures." The doyen of the art world was Sir Joshua Reynolds, recently knighted by the king, at the height of his fame and the Academy's first president. The Academy had only thirty-six members, and (after only two years in London) Kauffmann became one. Part of the reason might have been her relationship with Reynolds, rumored to be more than professional. Within one year Kauffmann was the most fashionable portraitist of the day, with commissions to paint Queen Charlotte and her children, Princess Augusta, and the king of Denmark. Four of her paintings were on show at the Royal Academy's first exhibition in 1769. The only cloud in her sky was that when the RA proposed her name as one of a small group of artists offering to paint the interior of St. Paul's Cathedral, Kauffmann was turned down on the grounds of her Catholicism.

The first Kauffmann picture to be exhibited in London, even before her arrival, had been one painted during her visit to Naples. It was a portrait of David Garrick, by this time the most famous actor-manager in England. He had originally come to London with his brother and set up in the wine trade. At one point he contracted to supply the Bedford Coffeehouse, a venue for literary and theatrical people. Garrick began writing for the stage, and in 1740 his *Lethe* had a successful run at Drury Lane Theater. In 1741 his *The Lying*

Valet was staged. The wine business was languishing, so he decided to make the theater his full-time career and became an actor, first appearing anonymously in *Richard III*. His performance astonished the audiences. Garrick introduced realism to the stage for the first time. He moved around the stage naturally, using a variety of facial expressions and speaking in a conversational tone. Alexander Pope[95] said 95 123 232
he would never be equaled. The Prince of Wales said he had not known what acting was until he saw Garrick. In both London and Dublin (where productions went at the end of the London season), Garrick was a box-office smash hit.

In 1747 he became actor-manager of the Drury Lane Theater and began to introduce changes. In 1762 he enlarged the theater, doubling its capacity. He cleared the stage of members of the audience, who had previously been in the habit of talking to the actors during performances. In the 1770s he hired John Philip de Loutherbourg, who revolutionized stage settings with the use of different levels to achieve perspective effect, lush backgrounds, transparent gauzes painted with scenes that would suddenly appear when lit, as well as free-standing pieces of stage furniture and colored lights. All of this made Garrick's productions the most exciting spectacles in London.

Before Garrick candles were the only source of illumination, either set on candelabra and lowered into a scene, or in the footlights. Garrick's new acting technique required better illumination, so he put reflectors behind the candles and added lighting from the wings. In 1785, after Garrick had left, the theater introduced an extraordinary new form of illumination that won instant plaudits: "The effect of this light, which is, in a manner, a new kind of artificial light, was brilliant beyond all expectation. The flame is bright without dazzling, strong and vivid, perfectly clear and yet at the same time steady, that the eye can bear to dwell upon it not only without pain, but even with some degree of peculiar satisfaction."

The new lamp had been invented by the Swiss Aimé Argand. Early in his adult life he lectured to the French Academy of Science on the distillation of spirits of wine and attracted the attention of winemakers in the South of France. By 1778 Argand was writing to the director general of the Herault Department (Jacques Necker,[96] soon to be 96 68 *141*

French director of finances and the father of the Romantic writer Germaine de Staël) offering information on his distillation process in return for a monopoly on the production of brandy and spirits of wine. In 1780 Argand demonstrated his technique in Montpellier to winemakers and two years later set up a large-scale distilling operation. Argand later said that it was during this period he began to think about his lamp. In 1783 he went to London to see if the lamp might have a British market. This made sense, since by now the country was in the early stages of the Industrial Revolution and factories needed better and safer illumination than candles. In his search for a manufacturer Argand went to Birmingham to the Soho
97 18 *40* works of James Watt and Matthew Boulton,[97] where a deal was
97 38 *68* struck. Boulton began production of the lamp in 1784.

The lamp itself consisted of a pedestal supporting a vase-shaped oil reservoir. Fitting on top of the reservoir was a metal structure consisting of two concentric brass tubes (pierced by ventilation slits), which held the circular wick and the wick-moving mechanism. A glass chimney rested on top of this. The advantages of the new lamp were that the entry of air through the slits past the wick and up through the open glass chimney facilitated the complete combustion of the fuel. The chimney ensured a bright, flicker-free light. In 1788, two British lighthouse towers, rebuilt on the southern tip of Portland Bill, were fitted with the new Argand lamps. By 1820 Argands were installed in some fifty British lighthouses. They would eventually be used in lighthouses all over the world. In some later versions the number of circular wicks was increased to ten. The lamp was ideal for lighthouses, since it gave off bright light and above all was less likely to cause fires (the main cause of lighthouse destruction), because there was no naked flame.

Lighthouse building reflected a growing demand for greater safety at sea following the general increase in the amount of shipping on the oceans, especially around Europe and across the Atlantic, carrying raw materials for the new Industrial Revolution factories and delivering their manufactured products. Argand's lamp had one unintended side-effect. Brighter lighthouses also made life easier for smugglers. Throughout the eighteenth century smuggling made up at least half of all English overseas trade. Smugglers dealt in almost every major commodity: tobacco, wool, tea, rum, brandy, wine, rice, molasses, slaves, logwood, flour, pitch, beef, pork, mercury, brass, ironware, cotton, canvas and nails. The attraction of smuggled goods was their low, untaxed price. It was this aspect of the trade that would lead to conflict between Britain and Spain.

Spain claimed a monopoly of all trade with her American colonies, but the Spanish economy was unable to supply the rapidly growing number of South American colonists with the goods they required. Early in the eighteenth century, of the twenty-seven thousand tons of merchandise legally despatched to Spanish America only fifteen hundred originated in the Iberian Peninsula. The rest was supplied by France, England and Holland. By 1731 the Spanish situation had worsened and smugglers stepped into the breach. Their technique was relatively simple. Regular ships from South America, bound for Spain with cargoes of gold and silver, cochineal, cocoa, sarsaparilla, balsam, indigo, dyewoods, tallow, vicuna wool and drugs, would leave Havana, their final port before the Atlantic crossing. Over the horizon they would be met by smugglers, carrying with them all the goods the colonists wanted but could not obtain from Spain, to be bought at untaxed prices paid for by some of the bullion on board the Spanish ships. To try to stop this trade the Havana authorities set up the first *guardacostas* (coastguards). These tended to be hired privateers working on the basis of "no prize, no pay," so they were often unscrupulous and violent.

In 1731 an incident occurred that was to have far-reaching consequences. The British brig *Rebecca,* sailing from Jamaica to London, was intercepted by Havana coastguards and boarded by a particularly violent individual, Juan de Leon Fandino. In the melee that followed the captain of the *Rebecca,* Robert Jenkins, had an ear cut off. On return to England his case for compensation languished in the courts. In 1738 it was revived and Jenkins was brought to testify before a House of Commons committee. At the hearing Jenkins produced a box in which he said was his ear. Much political capital was made out of the incident and public opinion was so outraged that in 1739 England declared war on Spain. It was a war that would become known as the "War of Jenkins' Ear."

When the war broke out a British naval captain, Lord George Anson, was recalled from Barbados where his mission had been to protect British shipping from attacks such as that on the *Rebecca.* Anson was to go on to become an admiral and to bring a number of reforms to the navy, including the classification of warships into six rates, the establishment of the marines, and the introduction of the blue-and-white officers' uniform. Meanwhile, in 1740 Anson was ordered to take six ships and fifteen hundred men to sail around Cape Horn into the Pacific and harry Spanish merchant shipping. Four years later he returned with so much plunder that thirty wagons were needed to haul it to the Tower of London for safekeeping. Anson's haul included 1,313,842 gold pieces-of-eight and 35,682 ounces of silver. Each crewman's share was well in excess of what he might have expected on departure from Britain, because Anson returned with only one ship and 145 men. The rest had perished not from their encounters with the Spanish but from scurvy. At the very beginning of the voyage, by the time they had reached landfall on the other side of the Atlantic, 200 were already dead. A year later only 323 men were left. The last would die in the third year of the voyage, leaving only enough to crew one ship.

The major problem with scurvy was that it weakened sufferers and left them vulnerable to other diseases. One of Anson's doctors wrote: "A most extraordinary circumstance—that the scars of old wounds, healed for many years, were forced open again; also many of our people, though confined to their hammocks, ate and drank heartily and were cheerful, yet having resolved to get out of their hammock, died before they could well reach the deck. It was also no uncommon thing for those who were able to walk the deck, to drop down in an instant on any endeavours to act with their utmost vigor."

The man who discovered the cure for scurvy did so as the result of work carried out only three years after Anson's return. His name was James Lind, and at the age of fifteen in Edinburgh he had been apprenticed to a surgeon. At the beginning of the War of Jenkins' Ear the twenty-three-year-old Lind joined the navy as a surgeon's mate. In the English Channel, aboard HMS *Salisbury*, Lind carried out what was probably the first controlled clinical trial in nutrition. For two weeks he kept a group of twelve scurvy patients on the same diet: breakfast of gruel with sugar; dinner of fresh mutton broth and pudding; supper of barley and raisins, rice and currants. Six pairs of the men were each allocated a different daily supplemental diet: a quart of cider, twenty-five drops of elixir vitriol, six spoonfuls of vinegar, half a pint of sea water, two oranges and one lemon, and a medicinal paste made of garlic, mustard seed, balsam, dried radish root and myrrh. By the sixth day of the trial the men on the citrus diet had much improved. The others continued to decline.

In 1748 Lind left the navy, returned to Edinburgh to an honorary medical degree and set out to write a short paper on his experiment. His work ended five years later with the publication of a four-hundred-page *Treatise of the Scurvy*, dedicated to Anson. In the long run the navy would react to Lind's report by introducing a supply of lemon juice for every Royal Navy ship. In the nineteenth century the practice was extended to the merchant marine, where their use of limes earned the sailors the nickname "Limeys."

When Lind received his degree the leading medical figure at Edinburgh University was Alexander Monro, professor of anatomy at the new Faculty of Medicine founded there only a few decades earlier. Monro (whose son was to treat Flora Macdonald[98] on her return **98** 83 *157*
from North Carolina) had impressive credentials, having studied in London, Paris, and in Holland at the University of Leiden, where he took lessons from the famous chemist Hermann Boerhaave. The large number of students attracted to Monro's dissection classes persuaded the university to come to an agreement with the city magistrates to supply the corpses of foundlings, stillborns, suicides, those who had suffered a violent death and those hanged for crimes. Unfortunately, supply failed to keep up with demand and led to the practice of grave-robbing. As long as the grave-robbing students left behind a corpse's shroud the removal of the body itself was not a criminal offense, although the activity caused such public outcry that Monro's windows were broken by a mob.

At the end of 1726 Monro published a major work, *The Anatomy of the Human Bones,* the first proper anatomy textbook, which contained minutely detailed descriptions. The book went into eleven editions and was translated into most European languages. In it Monro noted among other things that different nationalities could be identified by the shape of their cranium, that a man's stature decreases as evening approaches, and that the bone at a healed fracture is stronger than before the fracture. The extraordinary thing about the book was that it contained no illustrations. This was due to the fact that Monro's old teacher in London, William Cheselden (who got Monro elected to the Royal Society, was the queen's surgeon, friend of the great, famous for his fifty-four-second-long gallstone operations, and master of the Company of Surgeons) was himself planning to produce a book on bones, titled *Osteographica.* Cheselden's book was to have illustrations.

Osteographica was published in 1735 and ran to thirteen editions.
The title page carried an illustration of the technique Cheselden had
99 41 *76* used to produce his drawings. It showed a camera obscura.[99] At the
time, this consisted of a box, set into one side of which was a small
lens. This produced an inverted image (of any object at which the
lens was pointed) on the inside of the box opposite the lens. If the
image were made to fall onto translucent paper or a thick ground
glass surface it could then be traced. The technique made *Osteo-*
graphica one of the most accurately illustrated books ever published
on the subject of bones. The camera obscura had been originally
given its name by a short-sighted German astronomer named Jo-
100 42 *76* hannes Kepler,[100] who had first used it in July 1600 to draw a partial
100 108 *207* solar eclipse in the market square of the Austrian city of Graz, where
Kepler taught mathematics and astronomy.

In 1595, a year after his arrival in Graz, Kepler was struck by an idea so powerful he said he felt as if he had discovered the secret of the universe. The idea was not a new one, but it would lead to one of the most fundamental of all astronomical discoveries. Kepler was wondering why there were only six planets (known at the time) instead of twenty or a hundred, when he realized that this must have to do with what was known as the "five perfect solids." This idea struck him in the classroom, where he had just drawn a circle in a triangle in a circle. Kepler realized that the ratio of the two circles was about the same as that of the orbits of Jupiter and Saturn. He looked for more geometrical figures that might generate other orbits and hit on the "five perfect solids." These were classical Greek solid geometric figures, each of whose faces were identical: the tetrahedron, the cube, the octahedron, the dodecahedron and the isocahedron.

Each one of these figures could be placed in a sphere and all of its corners would touch the inside of the sphere. Kepler applied the figures to the orbits of the planets. Inside a sphere representing the orbit of Saturn he inscribed a cube; inside that, another sphere (the orbit of Jupiter), inside which he put a tetrahedron; then the sphere/orbit of Mars; then a dodecahedron inside which he put the sphere/orbit of Earth; then an isocahedron, then the sphere/orbit of Venus; then finally an octahedron containing the sphere/orbit of Mercury. This explained why there were only six planets: their orbits fitted into the five perfect solids.

In 1597 Kepler published his amazing discovery in his first book, *Mysterium Cosmographicum*. Kepler was now obliged to check observational data to verify his grand theory, and this data immediately threw up an awkward discrepancy. The orbits of the planets were not circles but ellipses and these did not fit inside the perfect solids. The data also showed Kepler that the planets traveled their elliptical orbits slowly when distant from the sun and faster when closer to it. Also, each planet moved more slowly the farther it was from the sun. Why was this so? Kepler made the great leap of imagination and proposed that the sun released a force of some kind that drove the planets around in their orbits. Since the force might diminish as did light with distance it would drive the planets less powerfully (and therefore less quickly) the farther from the sun they were. In spite of this discovery Kepler remained rooted in medieval cosmology, calling the mysterious solar emanation a "Holy Spirit Force" that acted like a lash to spur the planets on. Kepler then went on to measure the "force" and produced his three great laws: that the planets travel around the sun in ellipses; that a planet does not move at a uniform rate but in such a manner that if a line were drawn from the planet to the sun it would sweep over equal areas in equal times; and that the squares of the periods of orbit of any two planets are as the cubes of their mean distances from the sun.

At the end of September 1619 Kepler was working as provincial mathematician in the Austrian city of Linz when he was visited by a passing group of Englishmen headed by the earl of Doncaster, who was on his way to see the Holy Roman Emperor. The earl was accompanied by a chaplain who had read Kepler's work. The chaplain's name was John Donne, and one of his greatest poems had been written in reaction to the new cosmology. In the poem he wrote the now-famous lines:

> (the) new Philosophy calls all in doubt,
> The Element of fire is quite put out;
> The Sun is lost, and th'earth, and no man's wit
> Can well direct him, where to look for it.
> And freely men confess, that this world's spent,
> When in the Planets, and the Firmament
> They seek so many new; they see that this
> Is crumbled out again to his Atomies.
> 'Tis all in pieces, all coherence gone.

At the meeting in Linz Kepler is said to have given Donne a copy of his new book, *Harmonice Mundi,* to take back to King James I of England, to whom he had dedicated the work.

Donne's father was a well-to-do merchant and a member of the Ironmongers' Company. The Donne family was Catholic during a time of persecution, so Donne left Oxford without taking his degree, since this would have meant swearing the Oath of Supremacy and recognizing the queen rather than the pope as supreme authority over the church. Eventually Donne converted, took Holy Orders in the Protestant church, found powerful patrons and began a brilliant career as a preacher. By 1616 he was also a member of Parliament. In 1621 he was appointed dean of St. Paul's Cathedral in London, where he began a series of sermons that made him nationally renowned. Huge crowds pressed into the cathedral every time he spoke.

In the 1620s Donne met one of his parishioners, Isaac Walton, who kept a linen-draper's shop and was a vestryman at one of the several churches for which Donne was responsible. Like Donne's father, Walton was also a freeman of the Ironmongers' Company. The two men became friends, and in 1626 Donne officiated at Walton's marriage to Rachel Floud. In 1631, when Donne died, Walton was present at the deathbed and nine years later wrote his *Life of Donne*.

When the English Civil War came in 1642 Walton was a fervent supporter of the losing Royalist faction. In 1649 Charles I was beheaded and the eleven-year period of the Cromwellian Commonwealth began. Those who had supported the king were imprisoned, hanged or had their property and (in the case of the clergy) livings confiscated. In 1653 Walton published the book that would make him famous, *The Compleat Angler,* aimed at the many defrocked clergy of Walton's acquaintance. The book was written partly to provide the out-of-work churchmen with recreation during their time of enforced idleness and partly because fishing would provide food for those of them now suffering from poverty. As Walton noted, angling was a fit activity for clergy since the Apostles themselves had been fishermen and the task of priests was to fish for souls.

The plot of *The Compleat Angler* takes the form of a journey by two companions from London along the River Ware and back again. The book gives instructions for catching trout, salmon, chub, grayling, pike, carp, bream, tench, barbel, gudgeon and many more. Walton added entertainment for anglers in the form of poems, songs with music, arithmetic puzzles, dramas, anecdotes and proverbs. In the fifth edition Walton added a chapter on fly-fishing, written by his friend and fellow-angler Charles Cotton. Cotton was the son of a rich landowner in Derbyshire, attended Cambridge University, and traveled in Italy and France. He greatly admired Walton, who had also been his father's friend, and the two of them often fished in Derbyshire on the River Dove, where Cotton built a small fishing house for them to use. It still stands today and above the entrance can be seen the entwined initials "CC" and "IW." Cotton's comfortable lifestyle gave him time to write poetry and translate from the French. In 1671 he produced an English version of Corneille's *Horace* and in 1685 his last work was an English edition of the *Essays* of Montaigne, which still ranks as a masterpiece of translation.

Michel Eyquem de Montaigne, born in 1533, spent thirteen years as a magistrate in the court of Bordeaux. In 1570 at the age of thirty-seven he sold his magistrate's post and retired to his country estate where he lived in his library on the third floor of a round tower, leaving only for brief trips to Switzerland, Italy and Germany, and to serve two terms as mayor of Bordeaux. Montaigne's major contribution to European thought was to revive and make popular the ancient skeptic philosophy. On the beams of his study ceiling was carved the phrase, "All that is certain is that nothing is certain." The period in which Montaigne lived was fertile ground for the skeptic. His generation faced a problem that was new, urgent and fraught with danger: which form of Christianity was right, Protestant or Catholic? Each faith questioned the beliefs of the other. Protestants rejected the edicts of Rome and Catholics cast doubt on the literal interpretation of the Bible.

Montaigne's *Essays,* powerful exercises in skepticism, are unusually modern in their view. Montaigne was strikingly free from ethnocentrism and intensely interested in other cultures. He wrote about the newly discovered Indians of Brazil, giving a detailed description of their culture, and while remarking that they had "no trade, no knowledge of writing, no arithmetic, no magistrate, no political subordination . . . no riches or poverty, no contracts, no inheritance . . . no clothes, no agriculture, no metal," he refused to call them barbarians or savages, saying: "Everyone calls barbarous whatever is not customary with him." In an age rediscovering the Classics and discovering new parts of the globe Montaigne recommended the value of travel as a means of broadening the mind. Everywhere he went Montaigne sought to understand the local culture. In Switzerland he quizzed the Lutherans, in Verona the Jews, in Rome the flagellants, in France he questioned women accused of witchcraft. Prefiguring modern social anthropology, he said: "Every custom has its function."

Montaigne's work had a profound effect on a poor, provincial French writer named Pierre Bayle, who was eventually driven from France because he was Protestant. In Geneva he took courses in philosophy at the Academy and in 1681 moved to Rotterdam. There in 1696 he produced his great three-volume *Historical and Critical Dictionary*. The work was essentially a collection of biographies of historical writers and thinkers, and while Bayle included Classical figures in large part he dealt with those of his own time: humanists, Protestant theologians, and such recent philosophers as Spinoza and Hobbes. The *Dictionary* was based on Bayle's belief that all knowledge should be open to scrutiny and not simply passed on unquestioned from generation to generation. The work became a powerful weapon in the fight against intolerance.

Early in his Rotterdam period, in 1684, Bayle had begun to edit a monthly newspaper: *The News of the Republic of Letters*. This work put him in touch with a network of correspondents all over Europe. Two years later he published a letter supposedly written from the East Indies about the wars in Borneo between two queens, Mreo and Eneuge. The thinly veiled anagrammatic satire on "Rome" and "Genève" (that is, Catholics and Protestants) was written by another skeptic, Bernard de Fontenelle. Fontenelle settled in Paris in 1687 and began a literary career that shot him to fame with opera librettos, histories, comedies and poems. That same year he produced his best-known work, the first example of a scholarly work aimed at the general public, titled *Conversations on the Plurality of Worlds*. The book was a landmark in the popularization of science, but it raised dangerous questions about the relativity of knowledge because it postulated a nongeocentric universe. Fontenelle discussed the possible existence of other planets like the Earth, beginning the book: "It seems that nothing ought to interest us more than knowing how this world we inhabit is made, and whether there are other worlds like it which may be inhabited also."

In 1727 Fontenelle put a foot wrong with his *Elements of Infinite Geometry*, a work dealing with a level of mathematics about which he was not entirely qualified to write. There were, it seemed, mistakes in his math that attracted the attention of a major mathematical mind

in Switzerland who left Fontenelle in no doubt as to the error of his ways. Johann Bernoulli came from a Basle family that over three generations produced no fewer than eight mathematicians. Johann was apparently quarrelsome, quick-tempered and brilliant, and was said to be the man who made calculus understandable to the common man. His range was prodigious, including physics, chemistry, astronomy, optics and mechanics. Much of his time was spent working on differential calculus and corresponding with Gottfried Leibnitz, its coinventor (Newton did the same, independently). As part of Bernoulli's work in experimental physics he also investigated the recently discovered phenomenon of "mercurial electricity," first ob-
101 136 *254* served by the French astronomer Jean Picard[101] when in 1675 he noticed that movement of the mercury in his barometer produced a glow. Bernoulli was unable to explain the matter.

The man who did so was an Englishman, Francis Hauksbee. In 1709 he was demonstrator of experiments for the Royal Society under its president, Isaac Newton. In 1705 he showed the Society a nine-inch diameter evacuated glass globe that he had fixed to a device that would make it spin. As it rotated he pressed his hand against the glass and a purple light appeared inside the globe, so bright "that words in capital letters became legible by it." Hauksbee then experimented by rubbing evacuated glass tubes. These glowed, crackled and attracted small pieces of brass leaf, threads and wool. In 1709 Hauksbee published his findings, describing the glow and crackling noises as being akin to lightning and using the term "electricity."

After 1708 Hauksbee turned his attention to the phenomenon of capillarity: the way in which when a fine tube was dipped into a liquid the liquid crept up the tube. The narrower the tube the higher the liquid climbed. This also happened when the tubes were placed in a vacuum. Partly because of his link with Newton, and given the overpowering effect that the concept of gravity was having on all aspects of science, Hauksbee was convinced capillary action was related in some way to attraction. The particles at the surface of the liquid were clearly being attracted upward by the particles in the glass. Newton supported this theory in one of his 1717 *Optics* essays. This was enough for a London vicar named Stephen Hales, who in 1727 was investigating the behavior of sap in plants. In his *Vegetable Staticks* published that year, Hales, who had been sticking glass tubes into plants and watching how sap rose into the tubes, claimed that capillary action was at work in plants and might even account for the behavior of blood in human veins.

In 1740 Hales was stimulated by news of a typhus epidemic among sailors on board ship off Spithead awaiting transportation to America to take up the study of ventilation. A year later he published *A Description of Ventilators,* which included the design of a ventilator he had invented for placing in the outside wall of a building. Two pairs of large bellows were worked by a centrally pivoted horizontal lever. The bellows were provided with inlet and outlet valves so that as the lever was raised at one end one bellows drew air in from the outside. When the lever was depressed the other bellows forced the air into the room.

Hales tested his machine in a granary close to his church in Teddington, though its real purpose was initially for use in ships. There was a high incidence of fever among sailors and this was thought to be due to "disease-causing" air. The ventilator would replace this "effluvium" with "good" air. Eventually, in 1756 ventilators were installed in navy ships and in prisons, but not before Hales had also installed them in two hospitals, one in central London and the other a Middlesex smallpox hospital of which Hales was a governor.

Ventilation proved not to be the cure for smallpox, a prevalent and generally fatal disease. Forty years later an English country doctor, Edward Jenner, found the cure. He had begun his medical career apprenticed to a country surgeon for six years, then enrolled at the anatomy school in London run by the two famous surgeon-anatomists, John and William Hunter. After two years Jenner, now aged twenty-three, returned to his home village of Berkeley in Gloucestershire and set up in practice. In 1796 he came across several cases of cowpox contracted by milkmaids during the milking process. The usual symptoms were the appearance of large pustules

accompanied by fever. Jenner knew about the Turkish habit of inoc-
102 121 *224* ulation[102] against smallpox with the pus from smallpox pustules. This tended to act prophylactically as a preventative against contracting the disease but it was an unpleasant and sometimes fatal procedure. Jenner experimentally injected the eight-year-old son of one of his workers with pus from cowpox pustules and then infected the boy with smallpox. No smallpox symptoms appeared. Jenner called his technique (after the Latin word for cow, *vacca*) "vaccination." It was an instant success, and by 1800 the practice had spread throughout Europe and had reached America.

Most of the rest of Jenner's life was spent quietly in Gloucestershire, where he indulged himself in his passion for cuckoos. In 1788 he had been made a Fellow of the Royal Society following his publication *Observations on the Natural History of the Cuckoo.* In this he explained how cuckoo chicks hijack the nests of other birds using a hollow in their back to scoop up and tip out the eggs of the occupant. The chick would also do this to eggs containing cuckoo sib-

Fig. 22: *Not everybody greeted vaccination with the same degree of enthusiasm. A cartoon of 1802.*

lings. Just before his death in 1823 Jenner's last publication, on the migration of birds, was one of the first proper ornithological studies.

Jenner missed by three years the arrival of an American who would, so to speak, put ornithology on the coffee tables of the world. He was John James Audubon, the first great bird painter. Audubon had a varied career. Born in Haiti and brought up in France, he was sent in 1803 to Pennsylvania where his family owned a farm. In 1807, aged twenty-two, he set off to make his fortune in the West (at that time, Ohio and Kentucky). For thirteen years he moved from one financial disaster to another. The sawmill he opened failed, bankrupting him and an investor, the brother of the English poet John Keats. He tried running a store. It also failed. In Cincinnati he attempted taxidermy with some success. As he traveled he was also drawing and painting birds. In 1821 his fortunes changed when he landed the job of tutor to the daughter of a plantation family in Louisiana. In the five months that followed Audubon painted feverishly among the thousands of birds in the deep magnolia woods surrounding Bayou Sara, near St. Francisville.

When he had completed 435 paintings and failed to find a publisher, Audubon was persuaded to try England. Thanks to an influential contact, William Roscoe, within ten days of Audubon's arrival his drawings were on show at the Royal Institution. Audubon was an instant celebrity. Hundreds of people came to see the work, and he was lionized as the Romantic young woodsman from the New World. Audubon finally found a publisher in Edinburgh. By this time his plans for the work, to be titled *Birds of America*, had grown and he returned to America for more material. By now Audubon was an American national institution. Government schooners were placed at his disposal to take him to Labrador and other far-flung spots. In 1838 all five volumes of the great work were finally published.

In 1840 he received a letter from a seventeen-year-old boy named Spencer Fullerton Baird, describing a species Audubon had missed: the yellow-bellied flycatcher. Audubon invited the boy to visit and the two became friends. Ten years later the enthusiastic young naturalist had a specimen collection that filled two freight cars (2,500 American birds, 1,000 European birds, nests, eggs and reptiles, as well as 600 skulls and skeletons of American vertebrates and fossils). The freight cars carried the collection to the Smithsonian where Baird was to take up his appointment as assistant secretary.

The 1850s saw the beginning of the great age of exploration of the West, and Baird was responsible for shipping back to the Smithsonian all specimens collected by the explorers and surveyors. The information Baird gleaned from the numerous expedition reports was encyclopedic, so it was to him that the government eventually turned for information regarding the possible purchase of Alaska in 1866. Baird advised Congress to ratify the purchase.

Of the hundred or more expeditions in which Baird was involved perhaps the most exciting was that led by Ferdinand Vandiveer Hayden. In 1867 Baird obtained Hayden a grant from Nebraska for a survey of the state's geological resources. The survey was extended and refinanced over the next four years, and in 1871 Hayden returned from the Rockies with stupendous photographs by William Henry Jackson that persuaded Congress to declare the area a special place: the first National Park in the United States.

It became known, after the river running through it, as Yellowstone National Park. And it contained the biggest geyser in the Western world: "Old Faithful."

CHAPTER 8

Fire from the Sky

At the end of a bleak, wind-swept plain in the southern part of Iceland can be found one of the most extraordinary sights on the planet. Amid clouds of sulphurous steam and hot pools, every fifteen minutes *Strokkur* erupts, throwing a plume of scalding water seventy feet into the air. The word for this phenomenon, "geyser," comes from the old Icelandic word for "gush": *goysa.*

Geysers happen where the waters of an underground spring encounter hot rocks. This usually occurs in areas of volcanic activity and it happens in Iceland because of a gigantic, mile-long fissure known as "the thing that eats everybody": *allmannagja.* The fissure is part of a split in the Earth's surface that runs north-south the length of the Atlantic. As magma from the molten planetary core wells up through the split and hardens it forces the Atlantic tectonic plates apart at a rate of two centimeters a year. When the tectonic plates hit the continents around the edge of the oceans they are forced under the continental rim. The force of this impact causes buckling, throws up mountain ranges and on occasion causes cracks to appear. When this happens, magma is released to the surface by volcanic activity.

The first person to develop a theory to explain these planetwide disturbances was a German meteorologist named Alfred Wegener. Initially Wegener qualified as an astronomer but in 1905, deciding there were few astronomical discoveries left to be made (and anyway he lacked the advanced math), he took a job as technical assistant at the Aeronautic Observatory in Berlin and began meteorological work with weather balloons and kites. He spent much of the rest of his life collecting meteorological data on four expeditions to Iceland and Greenland.

It was after the first (1906–8) that Wegener was struck by the way the coastlines of the continents on either side of the Atlantic fit together. In 1858 Antonio Snider-Pellegrini had published a map showing this fit, and an American, Frank Taylor, had theorized that the Atlantic had split open and forced the continents apart. In 1911 Wegener came across paleontological evidence that strengthened his views. Almost identical fossil snails had been found in Africa and Brazil. There was no evidence of a now-sunken land bridge that might once have joined the two places, so in 1912 Wegener made public his continental drift theory. At a geological meeting he proposed that the harder rock of the continents rode like ships on the softer rock of the mantle and that the continents had drifted apart, perhaps impelled by the centrifugal force of the Earth's spin pushing the land masses away from the poles, or perhaps moved by currents in the molten magma beneath the surface.

Wegener's hypothesis was met with derision, considered by geologists to be the work of a meteorologist unqualified to speak about geological matters. It would take fifty years of further discovery to confirm Wegener's basic thesis. Until then most geologists thought Wegener was seeing things.

Ironically, his other major field of study was mirages. As part of his meteorological work he studied these phenomena, one of which, the *Fata Morgana,* was already being observed in the Middle Ages. The best early description of the most famous example, seen across the Straits of Messina between Sicily and Italy, was written in 1643 by an Italian cleric, Father Angelucci, looking across the straits towards Messina on the Sicilian side: "The ocean which washes the coast of Sicily rose up and looked like a dark mountain range." In front of the mountain "there quickly appeared a series of more than 10,000 pilasters which were a whitish-grey colour . . . the pilasters shrank to half their height and built arches like those of Roman aqueducts." Before it all vanished castles appeared above the aqueduct, each with towers and windows.

Mirages generally occur over flat surfaces like deserts and water and they are a complex product of air pressure and temperature, the temperature of the surface, the effect of gravity and the action of turbulence in the atmosphere. All these factors can combine to cause the light from a distant object to curve through the air and present a falsely positioned image to the observer's eye. In the *Fata Morgana* mirage the result is as Father Angelucci described, when the image can even originate in a flat object such as the sea and be transformed and distorted, defocused and moved in such a way as to create the famous shimmering "castles in the air" effect.

Fata Morgana is Latin for "Morgan the Witch." Morgan was one of the most powerful figures in medieval mythology and was thought to be the sister of King Arthur. Tales of the legendary sovereign of Camelot are generally reckoned to have originated in Wales with a sixth-century writer named Gildas. His work describes the fall of the late Romanized Celtic culture during the Anglo-Saxon invasions of England and records the Battle of Badon Hill, fought the year of Gildas's birth, at which Arthur is said to have fought. Scholars agree there was very probably a Welsh chieftain who fought the Anglo-Saxons invading Britain after the late fifth century and following the withdrawal of the Roman legions. In the early twelfth century the name of Arthur was appearing in the first *Histories of the Kings of Britain.* It is also at this time that the first mention of the Round Table appears. One interpretation of the legend is that it originates in Welsh mythology with Arthur as the sun god and the twelve knights representing the twelve months of the year. Be that as it may, by 1170, when a young French cleric named Chrétien de Troyes visited Glastonbury, in England (one of the places associated with the Arthurian legend), the stories of Camelot were well-known.

Chrétien appears to have been granted a benefice at the church of Saint Maclou in Bar-sur-Aube by Count Henry of Champagne, whose uncle Henry of Blois was abbot of Glastonbury. The abbot also had frequent contact with two of the authors of the *Histories,* William of

Malmesbury and Geoffrey of Monmouth, both of whom mention Arthur. So it may have been Chrétien's visit to England that provided him with material for his later poems about life at the Arthurian court and in particular the story of the "courtly love" affair between Lancelot and Arthur's wife, Guinevere.

In the courtly love relationship the man was the suitor, pleading his case in music and rhyme. The object of his affections would require him to undergo humiliations as proof of his love. The twelfth-century courtly love genre probably sprang from several sources. The first Crusades were bringing back from Constantinople stories of sensual delights, luxury and eroticism that had a powerful effect on people used to the deprivations of early medieval European life. At a time when marriage was primarily a matter of dynastic planning rather than passion, extramarital affairs were not unusual. Husbands often left wives for extended periods of time to go to war. Since adultery was a crime punishable by drowning, the platonic nature of the courtly love encounter may have been a means of sublimating sexual urges.

Perhaps the most intriguing origin of courtly love lies in the family relationships of Chrétien's employer Countess Marie, daughter of Eleanor of Aquitaine, who was herself the granddaughter of the first troubador, William IX of Aquitaine. The early twelfth-century troubador love-song tradition originated in an area of southwestern France which at the same time was in upheaval because of the activ-
103 51 *101* ity of a heretical sect known as the Cathars.[103] One of the Cathar beliefs was that marriage was an undesirable relationship, since it tended to produce children. The Cathars condoned extramarital sexual relations intended only for pleasure and not for procreation. Cathars also accorded women equal status with men at a time when a woman could normally only exercise her legal rights through her husband or male guardian. Before marriage she was the property of her father and after marriage that of her husband. Cathar attitudes toward the sexual and social position of women may have acted to trigger the development of courtly love.

The mystic Cathars were reformers who criticized the church for retaining worldly possessions and attacked churchmen for lax and immoral behavior. Cathars preached a return to the austere life of the early church Fathers. Cathar authorities, known as "the perfect ones," abstained from "impure" food (produce related in any way to procreation) such as meat, eggs and cheese. They also refused to kill any living thing. Sexual relations among *perfecti* were banned. They fasted on bread and water three times a week and observed special forty-day fasts three times a year. There were several reasons why the Cathar heresy attracted the particular attention of Rome. In the early twelfth century before the papacy had become centralized church authority was enforced through the local secular powers. In southwestern France many of these aristocrats were already Cathars.

Cathar beliefs struck at the heart of Catholic teaching because Cathars believed in two creators: God, good and perfect (the creator of the spirit), and Satan, evil and imperfect (the creator of the material world). Their argument was that a good Creator would not have made a material world so manifestly imperfect and full of evil. Cathars also believed that Jesus was merely an angel and that his human suffering and death were an illusion. Perhaps most important of all, the Cathars' appeal to the simple life and their attacks on the lifestyle of the Catholic clergy were winning converts among the poor.

For all these reasons Rome mounted a counterattack. In 1147 preaching missions by Alberic of Cluny, Geoffrey of Chartres and Saint Bernard failed. When Alberic preached in the cathedral at Albi (the center of Catharism) only thirty people turned up. Bernard was booed in the streets of Toulouse. So in 1198 Pope Innocent III declared the anti-Cathar Albigensian Crusade and then set up the Inquisition[104] to handle the long-term problem of heresy. The political **104** 50 *101*
attractions of a Crusade in the southwest of France were powerful. The king of France wanted to extend his influence south of Paris, plenary indulgences would be offered to those on the Crusade, and there would be none of the risks usually attached to Crusades in the Middle East. The Albigensian Crusade was ruthless and highly effective. Thousands of Cathars were burned without trial, their castles and churches destroyed, and their property confiscated. In one of the many massacres, at Beziers, when the papal legate was asked how

Fig. 23: *An early fourteenth-century illustration of the fate of the Cathars, denied even the* coup de grace *before burning.*

Crusaders would know Cathars from innocents he is said to have replied: "Kill them all. God will know his own." Fifty years later the Cathar heresy had been totally expunged.

The Crusade had also struck indiscriminately at Jews, who were numerous in the area. The Jewish intellectual community had many links with Cathars and may have influenced the mystical side of the Albigensian heresy. The southwest of France and northern Spain were centers of a Jewish mystic sect known as Cabalists, most of them intellectuals of the rabbinate who practiced meditation and the recitation of mystical formulae that would put the practitioner into an ecstatic trance during which he would see visions of heaven. The formulae for the chant varied according to different Cabalistic practices, but one of the most powerful techniques employed was that produced in the last quarter of the thirteenth century by an erudite Spanish Cabalist, Abraham Abulafia. Abulafia's *Path of Names* attached numerical values to letters of the Hebrew alphabet in such a way as to generate mystical relationships between words. The name-number equations showed, for example, that the number of the heavenly host was 301,655,172.

The technique of mystical word-number equations can most easily be described in English by giving the A–Z alphabet values from 1 to 26. The total letter-value number for the words "God is" (7 + 15 + 4 + 9 + 19 = 54) adds up to the same as that for "love" (12 + 15 + 22 + 5 = 54). Other mystical word relationships can be made in the same way. For instance, the number for "plague" is the same as "bad sky," suggesting a heavenly source of disease. "Holy Trinity" adds up to the same number as "Father, Son, Ghost." And "way of Cabal" is the same as "eternal peace." In 1274 Abulafia left Spain to travel and teach in Italy and Greece. Almost all his extant writings were completed in Italy, where he left his strongest and most lasting influence.

It was when a young aristocratic Italian scholar, Pico della Mirandola, became interested in Cabalism that Abulafia's ideas entered the mainstream of European thought and become a contributing factor in the development of modern science. Pico was Count of the small northern Italian town of Mirandola. Destined for an ecclesiastical career, in 1479 at the age of fourteen he was sent to study canon law at nearby Bologna University. Two years later he transferred to the University of Ferrara to study philosophy. Over the next four years he also visited Padua and Florence, meeting such major Renaissance intellectuals as Marsilio Ficino and Lorenzo de' Medici and being taught by the Jewish scholars Elia del Medigo and Flavio Mithridates, one of whom introduced him to the work of Abulafia and taught him Hebrew. Pico became obsessed with Cabalism and the Hebrew language. Between Abulafia's mystical numbers and the language of Israel Pico found a path to true faith. In his *Conclusions*, written in 1486, he risked death for heresy when he said: "There is no knowledge which makes us more certain of the divinity of Christ than magic and the Cabal."

Pico believed that the study of numbers would reveal truths about the universe and in introducing the idea into European thought Pico began the process that would end with the mathematical analysis at

the heart of science. Pico called this use of numbers "good magic," which would reveal relationships among all things in nature, stating: "By number, a way may be had for the investigation and understanding of everything possible to be known." *Conclusions* included a speech titled "Oration on the Dignity of Man," in which Pico took his belief in the ability to understand and control nature through numbers and turned it into the manifesto for the Renaissance. "Oration" foreshadowed what would become the scientific view of
105 86 *160* thinkers like Galileo.[105]

In 1490 in Florence Pico met a thirty-five-year-old, widely respected German scholar from the new University of Tubingen. Johannes Reuchlin was on the way to becoming one of the outstanding humanist scholars of the Renaissance. He was the first to teach Greek in Germany and had learned it from Greek refugees arriving in Europe after the fall of Constantinople in 1453. Reuchlin picked up a passion for Hebrew from Pico and took language lessons from Jacob Loans, a Jewish scholar who was private physician to the Holy Roman Emperor Ferdinand III. Over the next few years Reuchlin mastered the language and in 1506 produced the first manual of Hebrew grammar written by a Christian scholar. Reuchlin also took back to Germany Pico's fascination with the Cabal and the power of numbers. His *On the Cabalistic Art* was the first treatise on Cabalism by a non-Jew.

Like Pico, Reuchlin became convinced of the special value of Hebrew as a means of understanding Christian teaching. In 1508 he wrote: "I assure you, that no one of the Latins can expound the Old Testament unless he first becomes proficient in the language in which it was written. For the mediator between God and man was language, as we read in the Pentateuch; but not any language, only Hebrew, through which God wished his secrets to be made known to man." As Reuchlin's knowledge of Hebrew grew so did his belief in the need for the establishment of Hebrew faculties in European universities. This was a risky thing to suggest, given the anti-Semitic sentiment among many in the church. One of Reuchlin's opponents in this matter was a converted Jew, Johannes Pfefferkorn. In 1510 he asked Reuchlin to adjudicate in an appeal brought by the Jews of

Cologne against a court sentence ordering the burning of all Hebrew books. Reuchlin sprang to the Jews' defense and began a process that would finally bring him before a church court charged with heresy. The court proceedings dragged on for four years, and although Reuchlin was eventually acquitted, in 1520 he was ordered to pay the costs of litigation. Financially ruined and psychologically broken by the experience, two years later he died.

Reuchlin's defense in court had not been helped by a statement of support from his great-nephew Philipp Melanchthon, right-hand man to the church's greatest critic, Martin Luther. Melanchthon had first met Luther in 1518 when he arrived in Wittenberg, Germany (where Luther was a monk), to take up the post of professor of Greek at the newly established university. Luther, who only the year before had published the criticisms of Rome that would lead to the establishment of the Protestant church, attended Melanchthon's inaugural lecture, which was titled "The Improvement of Studies." It was a powerful, humanist appeal for a return to the original Aristotle and to the authority of the Bible. Melanchthon and Luther became instant friends. Both men shared the view that any reform of belief and practice had to start with education. A brief tour of Saxon schools convinced Melanchthon that root-and-branch change was urgently needed. He drew up a list of rules for the first-ever school inspectors. Schools were to be maintained by the civil authorities; teachers should be proficient in Latin and Greek; and schools were to be organized in three divisions: beginners, grammar-learners and advanced students. Melanchthon's rules also prescribed in detail the conduct of classes and the subjects to be taught. For the advanced class, these included Ovid, Virgil and Cicero, lessons in dialectic and rhetoric, prose composition, music and religious instruction. Melanchthon took the same modernizing approach to university instruction. He mounted an attack on the old Scholastic way of learning through disputation and worked out new statutes for faculties. This exercise was so successful that Melanchthon was called in to aid in the foundation of several new universities and to help reorganize older ones.

Not all Protestants (so called because they protested against Rome's negative reaction to Luther's theses) approved of Melanchthon's involvement with education. There was some criticism that his views were too liberal and that he had given way to Rome in many of his teachings. One of his particularly virulent critics was Andreas Osiander, who had been appointed professor of theology at the University of Konigsberg by the local duke, although not fully qualified for the job. Osiander issued pamphlet after pamphlet attacking Melanchthon for having forsaken true Lutheranism.

Osiander was interested in the mathematical sciences, and in 1540 when the first edition of Copernicus's first book (*Narratio Prima*) on the solar system was published, Osiander was sent a copy. The contents shocked him to the core. Osiander believed divine revelation was the sole source of truth, and the Bible did not describe the solar system as Copernicus did. The church taught the Aristotelian view of a cosmos with the Earth at the center and the sun and planets turning around it.

Copernicus's publisher was George Rheticus, a graduate of Wittenberg. When the full version of Copernicus's work was ready for publication in 1543, Rheticus was editing the manuscript in preparation for printing (at the Nuremberg press of Johann Petreius) when he was called away to the University of Leipzig to become professor of mathematics. He left the editing to Osiander, who first changed the title of the Copernican work from *On the Revolutions of Planets in the Sky* to the one by which is it still known, *On the Revolution of the Heavenly Bodies*. He also inserted a preface stating that while the author described a sun-centered system, the scheme was intended merely as a piece of mathematical convenience to make the work of astronomers easier and did not claim to reflect what was actually happening in the sky. This explained why Osiander had changed the title. The use of the word "planet" would have introduced the idea of orbiting bodies, which could include the Earth. Copernicus received a copy of the book with its new title and preface on his deathbed, when it was too late to do anything about it.

The next book Osiander received had been printed by the same Petreius Nuremberg press. This time the book was dedicated to Osiander himself. It was a text on algebra (*The Great Art*) written by an Italian physician friend of Osiander's, the inventor and gambler Girolamo Cardano. The illegitimate son of a ne'er-do-well mathematician and lawyer, Cardano first studied at the University of Pavia, then taught philosophy while gambling away his leisure hours and writing a book, *Games of Chance.* In it he expounded the first law of probability: that throwing dice involved the impossible (throwing a seven with a single die), the certain (that one side of the die must fall uppermost) and the probable (that the first throw might be a six). If the impossible were set as zero and the certain as one then all degrees of probability in between could be calculated in fractions (that is, the chances of throwing a six is one in six). Later in life Cardano invented the universal joint still in use in the transmission shaft of every car today and known as the Cardan shaft. In 1525 Cardano qualified in medicine at the University of Padua and began a career as a doctor while working on what would be his magnum opus—*The Great Art* algebra book dedicated to Osiander.

In 1551 Cardano received a letter from the personal physician to Archbishop John Hamilton of St. Andrews, Scotland, asking if he would consider visiting the archbishop in order to treat his asthma, which had become so serious the prelate was unable to move from his house in Edinburgh and was suffering twenty-four-hour-long attacks once a week. Hamilton was no ordinary churchman, being the illegitimate brother of the earl of Arran, regent of Scotland during the infancy of Mary Queen of Scots. On June 29, 1552, after a journey from Italy lasting over six months, Cardano arrived with a new treatment for asthma that he had developed over the few previous years and was anxious to test.

Standard treatment for all conditions at the time assumed that the cure for any disease lay in one of two different conditions of the brain. Some physicians believed a healthy brain was "hot," some "cold." Hamilton's personal physician had been of the "hot" persuasion and this had led him to prescribe a cure that would make Hamilton's brain hot. Treatment confined the patient to rooms heated to stifling point and permitted only scalding food and mulled wine, in order to keep him in a constant sweat. Since the archbishop's lifestyle was already overindulgent the treatment made things worse. Cardano took the "cold"-brain approach, recommending cold showers, walks before breakfast, plain and nourishing food, plenty of fresh air, lots of sleep, and periods of rest every morning ("but not with harlots"). Within two weeks Hamilton was so much better he persuaded Cardano to stay until September, whereupon Cardano departed for Italy much the richer, leaving a healthy churchman behind. Hamilton would not keep his health for long, thanks to the intrigue and turmoil that was to follow.

The contemporary political situation in Scotland was complicated by the fact that the country swung between Catholicism and Protes-
106 62 *128* tantism. In 1560 the Catholic Mary Queen of Scots,[106] for reasons of marriage absent from Scotland for twelve years (leaving the country in the hands of a regent), returned to the country on the death of her husband, the French king Francis. At this time Scotland was Protestant and Mary's Catholicism was accepted with the stipulation that she restrict her observance to her private quarters. Elizabeth, queen

Fig. 24: *A portrait of the young Mary Queen of Scots on her return to Scotland from France in 1560.*

of England, together with many of Mary's own advisers, now strongly advised a Protestant marriage, but contrary Mary (of whom the nursery rhyme speaks) decided instead on Henry, Lord Darnley, who was a Catholic. This raised Elizabeth's hackles, since through family ties both Mary and Darnley had claims to the English throne. Mary's grandmother had been Margaret Tudor, the elder sister of Elizabeth's father, Henry VIII. Darnley, Mary's cousin, was Margaret Tudor's grandson.

The marriage between Mary and Darnley was a disaster. Darnley proved to be a jealous, drunken lout who resented not being king of Scotland in his own right and who eventually connived at the murder of Mary's thirty-three-year-old Italian personal secretary, David Rizzio, claiming that the queen was having an affair with him. Mary retaliated by appointing the earl of Bothwell as her principal adviser. On the night of February 10, 1567, Darnley was killed in a huge explosion that wrecked the house in Edinburgh where he was recovering from syphilis. The conspirators were soon caught, having left behind them an easily traced empty powder barrel, bought candles at the last minute from a nearby shop and obtained fuses from soldiers of Darnley's own voluble guards. In 1571 Archbishop Hamilton would be hanged in his own vestments for complicity in the murder.

Meanwhile Bothwell divorced his wife and persuaded Mary to marry him. She was several months pregnant at the time, almost certainly by Bothwell. On May 15, 1567, Mary committed the ultimate folly, marrying Bothwell in a Protestant ceremony. This cost her the last vestiges of her political support in Europe. An army of Scottish nobles came against her and at the Battle of Carberry Hill, on June 15, she was captured. By the following year she had been taken across the border to an English prison and accused of treason for having claimed the English throne. Nineteen years of incarceration later, Elizabeth had Mary executed.

Meanwhile, Bothwell decided to flee the Scottish mainland. One of his titles (conferred on him only recently by Mary) was duke of the Orkney Islands, an archipelago well to the north of Scotland. Bothwell headed for Orkney with eight ships and a plan to set up "an empire of the sea." Unfortunately, the local Orkney sheriff decided discretion was better than valor and refused him entry. The pursuing Scottish ships caught up with Bothwell and he was forced to make a run for it, eventually heading across the North Sea. As he came within sight of the Norwegian coast Bothwell had the misfortune to be intercepted and escorted into port at Bergen. Bothwell revealed his identity as Duke of Orkney. Things were going well when bad luck intervened. When the investigation into Bothwell's arrival began, one of his old flames, Anna Throndsen, turned up. Seven years earlier she and Bothwell had met when he was in Denmark and she had run away with him, first to France and then to Scotland. After years of living as his unacknowledged mistress, when Bothwell divorced his wife Anna expected marriage. So when Bothwell married Mary Anna went home in a rage. Now it was payback time. Her allegations of seduction and injury were enough to cook Bothwell's goose. He was moved to Malmö prison for greater security, while discussions were opened with Scotland about what to do with him.

Bothwell was visited in Malmö by the venerable French ambassador to Denmark (which ruled Norway and Sweden at the time). His name was Charles de Dancey, and he offered to take a letter from Bothwell to the French king and to intercede on Bothwell's behalf with the Danish authorities. Although Dancey was well-respected in Denmark there was little he could do for Bothwell. With Mary in prison nobody wanted the runaway Bothwell back in Britain to foment trouble. He was moved to a remote and rigorous prison regime in Dragsholm, Zeeland, where eventually he went mad and died in 1578.

Meanwhile Dancey continued a busy social life in Denmark, part of which involved his close friendship with the Danish astronomer
107 135 *254* Tycho Brahe,[107] a well-heeled young aristocrat whose family were friends of the king. Free to travel and indulge his passion for astronomical instruments, after initial studies at the Universities of Copenhagen, Leipzig and Wittenberg Tycho visited the great instrument-making center of Augsburg and began the serious study of the stars. In Augsburg he commissioned the construction of the latest model of quadrant, marked with single minutes of a degree, which he began to use to measure the apparent motion of the plan-

ets. Because of his father's illness in 1571 Tycho returned to the family home at Knutsdorp, Sweden, and went to live with his uncle at the nearby Cistercian monastery of Herrevad, where the two men set up a chemistry lab and worked on techniques for making gold.

On November 11, 1572, Tycho was returning from the lab when he looked up and saw something impossible: a bright new star in the sky. It was a supernova, and it was impossible because according to the pope and Aristotle the heavens were unchanging, so there could be no new stars. For thirty days Tycho measured the new star's position relative its nearest constellation, Cassiopeia. From any angle their angle of separation remained constant, indicating that the star was out in deep space together with the constellation. In 1573 in Copenhagen Brahe pointed out the star to Dancey and later that year he published the book he had now written about it, titled *On the New Star.* The book and its potentially heretical contents made Tycho famous all over Europe and established him as an astronomer of the first rank. Three years later the king of Denmark had given him the island of Hven (between Denmark and Sweden) and Charles de Dancey was there to lay the foundation stone for Tycho's great observatory, Uraniborg.

At Uraniborg Brahe produced a set of updated star tables dedicated to the Holy Roman Emperor, Rudolph, who was so impressed he asked Tycho to come to Prague as imperial astronomer. He also invited a young German astronomer, Johannes Kepler,[108] to be Brahe's **108** 42 *76*
assistant. A year later Tycho was dead, buried together with the metal **108** 100 *182*
replacement nose that he had worn ever since a duel in his university days. The existence of the nose is only known because it was mentioned by Tycho's assistant on Hven.

This was a mathematical young Dutchman named Willem Blaeu, who returned to Amsterdam in 1596 after his work with Tycho and turned his astronomical experience to good use by setting up a company dedicated to printing navigational data (for which a knowledge of stars and star tables was essential) and maps. With a shop on the Amsterdam waterfront, Blaeu was able to pick up the latest information from returning sea captains and keep his maps and tables up-to-date. By 1633 he had one of the most successful mapmaking houses

in Europe and had published charts of the four known continents, as well as books on cosmography, hydrography, topography and his own *Light of Navigation* manual for navigators. That year he was ap-
109 44 *80* pointed official cartographer to the Dutch East India Company,[109] the
109 66 *137* greatest exploration and trading organization in Europe. He was to hold the job until his death in 1638.

Improved navigation and better maps were critically important to the East India Company, because cargoes from the East could be re-exported throughout Europe for profits of as high as 600 percent. Since the time of the company's establishment in 1602, more and more ships returned with luxuries for which Europeans would pay high prices: dyestuffs, pepper, silk, porcelain, tea, saltpeter, cinnamon, borax, musk and sugar. Voyages in search of these commodities were made easier by one of the cartographers whose maps Blaeu printed. He was a German named Gerard Kremer, also known as Mercator. Mercator solved a basic problem facing every navigator. Steering a straight-line course across latitude (north-south) lines that curved to converge at the pole meant that at the crossing of each line the compass bearing would change. In 1569 Mercator projected the globe onto a cylinder, which meant that the latitude and longitude (east-west) lines now crossed at right angles. This in turn meant that a straight-line course crossed all latitude lines at the same angle, making the navigator's life simpler. The fact that this map distorted the size of high-latitude locations such as Greenland was considered relatively unimportant. These were not the places to which commercial ventures went in search of profit.

In 1674 the first complete collection of charts to use Mercator's technique was produced by an Englishman, Robert Dudley, who titled the work *Dell'Arcano del Mare* (On the Secrets of the Sea). It was
110 85 *160* created for and dedicated to his employer, Ferdinando II,[110] duke of Tuscany. By the time of publication Dudley had lived in Italy for about forty years, ever since running away from England with his mistress, leaving wife and family behind. Dudley was the illegitimate son of the earl of Leicester, and since his mother's marriage to the earl had been secret and after her death the earl had married again,

family feuds put Dudley's inheritance in question. After his arrival in Italy, Dudley illegally assumed the title of Earl of Warwick, as a result of which all his English estates were confiscated. With his boats burned, he turned to boat-building, and impressed the Tuscan duke with his maritime knowledge (Dudley had spent several years with the English Navy). Dudley was given control of the shipyards at Pisa and Livorno, arranged for English ship-builders to work there and began to build warships for the duke. Dudley was also given the tasks of draining the marshes between Pisa and the sea, installing a fresh-water aqueduct to supply Pisa and building a canal linking Pisa and Livorno. He also persuaded the duke to declare the newly rebuilt city of Livorno (Dudley designed the harbor mole) a free port and "a place of universal toleration," thus attracting religious refugees of every stripe from all over Europe and vastly increasing the duke's fiscal revenue as Livorno rapidly became a highly profitable major international entrepôt.

Dudley's predecessor in water-management (and possibly, for a brief time after Dudley's arrival in Italy, his boss) was the celebrated engineer and architect Bernardo Buontalenti, who worked for the Medici family for over sixty years. Of his many architectural projects, perhaps the most famous are the Belvedere Fortress in Florence, the Villa Pratolino, and the fortifications of Livorno. But Buontalenti's special talent was with machinery. He began by designing water-lifting devices and complex water gardens with fountains and grottoes featuring water-powered automata. By 1589 he was in charge of the spectacular entertainments staged by the Medici on the occasion of a marriage or the arrival of a special visitor. In one case Buontalenti flooded the Pitti Palace courtyard to a depth of five feet for a mock sea battle. Buontalenti also designed fire-breathing dragons, exploding volcanoes, moving clouds on which gods were transported, collapsing castles, mountains, rocks and trees that rose and sank through the floor and thunder effects created offstage by rolling cannon balls down metal tubes. The plays that were presented included intervals, and in these short breaks a new musical form began to appear in which singers and dancers would be accompanied by musical instruments as they enacted scenes to the accompaniment of Buontalenti's spectacular effects. Out of these early efforts came the first true opera, Peri's *Dafne*, staged in 1598.

Buontalenti's scenography was radically improved by another Italian who worked in Venice, named Giacomo Torelli. His contribution to stagecraft and design would remain standard procedure well into the nineteenth century. Torelli's use of machinery may have originated in 1640 during his brief period at the Venice shipyards, where the shipbuilders used a considerable amount of automation, with machinery operated by rope and pulley. Torelli's theatrical innovation involved cutting diagonal slits on either side of the stage floor. Through these slits Torelli erected small poles that supported scenery. Beneath the stage the poles were mounted on wheeled trolleys running on rails. The trolleys were operated by ropes wound onto a central drum. When the drum was turned with a counterweight all the trolleys would move at once, causing scenery to come on or off the stage with comparative rapidity. In 1645 the technique
111 64 *135* impressed an English visitor, John Evelyn,[111] who reported that in one of Torelli's plays the scenery changed no fewer than thirteen times. That same year Torelli was summoned to France by Giulio Mazarin, the Italian minister running the country for Louis XIII. At Mazarin's request Torelli introduced the French to what became known as the "theater of machines" and started a vogue for plays that included spectacular effects the machines made possible.

Giulio Mazarin was so popular he survived to serve under the next king, Louis XIV. A well-educated man, Mazarin had a passion for art and literature and provided pensions for Racine, Molière, Corneille and others. Mazarin's pride and joy was his collection of over forty thousand books, acknowledged as the best library in Europe. The books had been collected for Mazarin by his librarian, Gabriel Naudé, who in 1627 had published the first proper study of library science in a book titled *Advice on Establishing a Library,* which included instructions on how to choose, classify and arrange books, as well as hints on the decoration of a library, the treatment of staff and even on how to dust the books. In 1661 Naudé's book was translated into English by the same John Evelyn who had seen Torelli's play in

Venice. Evelyn then became an expert in library science and set up his own collection of books organized according to Naudé's principles.

One of Evelyn's friends was also interested in using Naudé's information to help him build a library of books that would help him with his grand scheme to reform the English navy. This maritime bibliophile was Samuel Pepys, and he had met Evelyn because of the latter's connection with navy hospitals. By 1685 Pepys was secretary to the navy and responsible only to the king. Pepys's reforms effectively founded the modern British navy and dealt with virtually every aspect of naval life on shore and at sea. He introduced formal training for officers, who would now be required to be able to navigate. Discipline, pay, pensions and medical treatment were regulated. Uniforms and saluting were formalized. Dockyards were modernized and the whole system of tendering for naval contracts was properly established. On board ship, guns and orndance were standardized and the number of a ship's complement now related to the size of the vessel.

Fig. 25: *Samuel Pepys, pictured with the manuscript of his song "Beauty Retire."*

Pepys persuaded the government to agree to supply the fleet with six months' stores in advance. Above all, he instigated the country's largest-ever ship-building program to ensure English supremacy at sea for generations.

The only aspect of naval life that Pepys did little to reform was the practice of signaling, which in the mid-seventeenth century was a primitive procedure. If a ship needed a delivery of wood it hung up an axe. A request to come to dinner was signaled by hoisting a tablecloth. In 1673 the first proper signal book appeared with colored drawings of fifteen flags and each of the positions to which they should be hoisted. By 1782 the number of flags had increased to fifty. By 1799 various patterns of three or four flags made it possible to send up to 340 signals and the signal book also included eighty last-minute manuscript additions. However, by this time the signaling problem was most acute on land. Communications between the Admiralty in London and the various naval ports had for centuries been done by courier at the speed of a horse. Delivery could take days, or in the case of foreign stations, months.

112 10 *34* This problem was solved by Britain's French enemy, Napoleon,[112]
112 43 *78* in 1792. Surrounded on all sides by the Allies, Napoleon was desper-
112 59 *120* ate for a better way to communicate with his widely dispersed
armies. On March 22 that year, the French Legislative Assembly was given a demonstration of a new communications system invented by a priest named Claude Chappe. Chappe's system involved sending signals by means of a twelve-foot horizontal wooden arm pivoted on a vertical beam on top of a tower. At each end of the horizontal beam was another three-foot beam, also pivoted. By means of pulleys and ropes all the movable beams could be made to take up a large-enough number of configurations to send a significant number of signals. These could be seen through a telescope by an observer on another distant, similarly equipped tower, who would then relay the message by the same means to the next tower, and so on. By 1794 the system was in operation over 210 kilometers, reporting in one instance on a military event only an hour after it had taken place instead of the ten hours it would have taken by conventional means.

That year a copy of Chappe's drawings happened to be in the possession of a French soldier captured by the British. The drawings
then fell into the hands of the Reverend John Gamble,[113] a British 113 19 42
army chaplain, who promptly made a number of improvements and sent drawings to the Admiralty. To Gamble's chagrin, Chappe's idea had also reached the ears of Lord George Murray, who had also made improvements. Since Murray was a well-placed aristocrat it was his version the Admiralty proposed to adopt. In 1795, after successful trials on Wimbledon Common outside London, several chains of the new "semaphore" stations were set up.

By 1805 one of the chains extended as far as the major naval port of Plymouth, and a contemporary described the arrival of a message: "A single signal has been transmitted to Plymouth and back . . . [to London] in three minutes, which by the telegraph route is at least 500 miles. In this instance, however, notice had been given to make ready and every captain was at his post to receive and return the signals. The progress was at the rate of 170 miles in a minute, or three miles per second, or three seconds at each station; a rapidity truly wonderful."

Because of his work in prisoner-of-war exchange Gamble had also been able to facilitate the movement of people and documents through the English Channel ports, and it was thanks to this that in 1810 the patent for another French invention arrived in England. It was for a process devised by a champagne-bottler named Nicholas Appert for boiling food in a hermetically sealed bottle. This killed the bacteria in the food and kept it from putrefying for several months. The preserved food had tested by the French navy and was a great success.

When Appert's patent arrived in England it ended up in the possession of a businesssman named Bryan Donkin, who had an interest in an iron works and who realized that preserved food might last better and longer in metal containers. A canning factory was established by Donkin and his partner, John Hall, in 1812. By 1818 the partners were producing cans of cured beef, boiled beef, carrots, mutton and vegetable stew, veal and soup.

That year canned food formed part of the provisions taken by the Ross expedition to the Davis Straits in northern Canada. The expedition, headed by Captain John Ross and including his nephew James, aimed at finding the Northwest Passage. They failed, and in 1829 Captain Ross took a second expedition (financed by Felix Booth, manufacturer of Booth's Gin), which failed once again to find the Passage. However, on June 1, 1831, John Ross's nephew James crossed the ice on foot and reached the Magnetic North Pole. He identified the exact spot by suspending a magnetized needle on a thread and watching it assume a nearly vertical position. He promptly raised a flag and claimed the location (70.5 N, 96.46 W) in the name of King William, in spite of the fact that since the Magnetic Pole moves it was probably no longer exactly there even as he claimed it.

James was now bitten by the Magnetic Pole bug, and in September 1839 he headed his own expedition to Antarctica in search of the South Magnetic Pole. In January 1841 the explorers were stopped about three thousand yards short of their target by a mountain range rising in some places to twelve thousand feet. However, other expedition objectives were more than adequately met, as the expedition also discovered Victoria Land, the Ross Sea, McMurdo Sound and the Ross Ice Barrier.

The assistant surgeon aboard Ross's ship was a young man named Joseph Hooker, son of the director of the Royal Botanic Gardens in Kew. In 1872, after his return to England, Hooker, by now director at Kew, received a shipment from Brazil of seventy thousand rubber
114 33 *66* tree seeds. Vociferously encouraged by Charles Macintosh,[114] who had discovered how to liquefy rubber and use it for waterproof clothing, Hooker set about attempting to grow seedlings from the seeds. Only 4 percent of them germinated, and eventually 1,919 plants were dispatched to the Botanical Gardens at Peradeniya, in Sri Lanka. A few plants were also sent to Singapore, where they did not survive. A few seedlings also went to Malaysia.

Several years later the Malaysian plants had done well enough for rubber seedlings to be planted in Java. In 1884 the first commercial rubber-tapping occurred in the Sri Lanka plantations and from then on the East became the prime source of British rubber. The industry rapidly expanded. By 1922, 85 percent of a total world rubber supply of nearly 380,000 tons came from the Eastern plantations.

During World War II the Japanese captured the plantations in the Malaysian archipelago and Sri Lanka remained the only source of Allied rubber supplies. This was a disastrous turn of events, since one of the principal uses for rubber in wartime was as a key ingredient in the manufacture of incendiary bombs. In July 1943 a rubber-benzol mixture (the rubber helped the benzol to burn more slowly and to stick to anything it touched) formed the main ingredient of three million incendiaries dropped on the German city of Hamburg, destroying the city and killing between forty thousand and fifty thousand people.

CHAPTER 9

Hit the Water

When Japanese forces took the Malaysian archipelago during World War II they left the Allies with an acute problem: how to run their war machine without plentiful supplies of rubber. For such things as tires and waterproof clothing essential to troops the problem was rapidly solved by the development of neoprene, an artificial rubber
115 58 *119* first discovered by an American chemist named Julius Nieuwland[115] and produced in an industrial process developed by DuPont.

This left one major problem unsolved: how to make incendiary bombs. These had previously used a mixture of rubber thickener, benzol and phosphorus, the rubber being used to retard the speed with which the benzol burned. A new thickener was needed. In 1940 when it began to look as if the war in Europe might turn into a prolonged conflict, the American government decided that science was likely to play a decisive role. The National Defense Research Committee was therefore established under Vannevar Bush, president of the Carnegie Institute. One NDRC division was to be dedicated to research into bombs, fuels, poison gases and chemical weapons. This division was placed under the command of James Conant, president of Harvard.

With the attack on Pearl Harbor in 1941 and the entry of America into the war, a Harvard professor, Louis Fieser, was asked to solve the rubber problem. The specifications were that the rubber substitute had to remain thick at temperatures as high as 150 degrees Fahrenheit (for use in the tropics) and not become brittle at minus 40 degrees (in a bomb bay at altitude). It had to be tough enough to withstand an explosive blast without shattering and it had to survive lengthy storage without deterioration. Most important, it had to be adaptable to a simple field-loading operation.

In July 1942 Fieser had his product. It was called "napalm." By the end of the war output was more than 70 million pounds a year, and a total of 33 million bombs had been made. Each bomb consisted of a rod of high explosive that exploded, shattering the napalm container and at the same time releasing phosphorus, which then ignited the napalm. In the early stages making a bomb was simple. Napalm powder was stirred into gasoline or benzol in an aircraft fuel drop-tank and left overnight to thicken. Explosive and phosphorus were then added. On impact the bomb created a fireball, which burned intensely for ten seconds, giving way to a fire of reduced intensity that burned for up to ten minutes over an elliptical area thirty by ninety yards. Napalm was terribly effective, and in 1972 after adverse publicity surrounding its use during the Vietnam War, the United Nations passed a resolution deploring its use.

One of the key ingredients in napalm was palm oil, which had been available in large quantities since as early as the 1820s, when it was first imported (from Indonesia, the Philippines, Malaysia and Sri Lanka) as an ingredient for soap manufacture. The man who made soap-making into an industry was Michel Chevreul, director of the Gobelins tapestry factory outside Paris. Chevreul was an expert on fats of all kinds, thanks to his interest in the behavior of animal fats in yarn. It was Chevreul who persuaded a young French chemist named Hippolyte Mèges Mouriès to study fats. Mèges had already achieved some success with various inventions and discoveries: a remedy for syphilis, a technique for using egg yolk in leather-tanning, a new way of making bread, and effervescent tablet manufacture. In 1852 he turned to fats.

One of Europe's major problems at the time was population increase due to industrialization and the fertilizer-enhanced growth of cereal crops. Between 1750 and 1850 the European population had grown from 140 million to 266 million. The majority of these were factory workers whose diet was nutritionally deficient, lacking in protein and fat to provide the energy they needed for work. By 1850 the fat supply was well below requirement. Beef suet would have filled the need, but it could not be spread. The demand for butter rose steeply and so did its price. Mèges was to resolve the impasse.

116 26 *50* On Napoleon III's[116] imperial farms at Vincennes Mèges pressed
116 126 *236* beef suet at forty degrees Centigrade and obtained a fat that melted at
116 129 *246* five degrees. Churning this with milk produced a material that
would spread on bread, which Mèges called "margarine." In 1871 he sold the patent to the Dutch firm of Jurgens, as well as to British, American and German manufacturers, and margarine went into production everywhere. Unfortunately supplies of beef suet soon ran out. A substitute had to be found that would be soft enough to spread but hard enough not to run. There was no such material available until 1897 when a pair of French chemists, Paul Sabatier and J. B. Senderens, discovered that vegetable oils were fluid because they had a lower hydrogen content than solid fats such as butter and lard. In 1902 a German, Wilhelm Normann, found a way to add extra hydrogen to oils. In its final industrial form the process involved pumping oil at around 180 degrees Fahrenheit into a closed vessel under pressurized hydrogen and then adding a catalyst consisting of fine particles of nickel deposited onto an inert powder called *kieselguhr.* The nickel catalyst caused hydrogen molecules to attach themselves to the oil molecules. As a result the eventual melting point of the oil could be determined (by how much hydrogenation took place), so that the hydrogenated oil would melt only at temperatures above that of normal use in margarine.

Kieselguhr is produced by grinding a friable sedimentary rock resembling chalk, and it has many uses besides hydrogenation, as an ingredient in toothpaste, ceramics, detergent, insulation and plastic. It is also employed as a filler in brick, paint and paper. One of its first uses was as the inert material holding nitroglycerine in sticks of dynamite. *Kieselguhr* is also known as "diatomaceous rock" because it is formed from the shells of diatoms, or plankton, which fall to the ocean floor when they die and over lengthy periods of time form sediments that then harden into rock.

This fact was discovered by a German researcher named Victor Hensen, from 1868 professor of physiology at the University of Kiel, who investigated the organ of hearing in grasshoppers' forelegs and went on to identify Hensen's duct in the human cochlea. Hensen's hobby was marine biology, and when he was a member of the Prussian Parliament he lobbied for funds for research programs that would benefit the German fishing industry. During this time it occurred to him that the most valuable contribution to fisheries would be one that assessed the productivity of the ocean itself. To this end Hensen decided to investigate the smallest organisms in the sea that might form the base of the food chain. In 1889, with the aid of specially modified silk millers' nets normally used to separate different grades of flour, Hensen set off on the great German Plankton Expedition to survey the whole of the North Atlantic.

Plankton (so-called from the Greek work for "drift") are tiny vegetable cells enclosed in a two-part shell, and they are the most prolific organisms in the sea. They are also ubiquitous, found in fresh and salt water, in mud or sand in shallow water or near the surface of the deep ocean. Plankton are so small that a jar of sea water contains millions of them. This meant that their numbers could only be assessed using statistical methods based on a representative microscopic count. The Plankton Expedition, on board the steamship *National,* lasted 115 days and criss-crossed the principal biogeographical zones of the North Atlantic as far apart as Greenland and Bermuda, the mouth of the Amazon and the Cape Verde Islands off the west coast of Africa.

Hensen made several significant discoveries during the expedition. Plankton exceed the mass of all other organisms in the sea. The high seas are generally poorer in plankton than are river mouths and coastlines. The deep-blue color of the ocean represents an almost complete lack of plankton. Most interesting and unexpected of all, Hensen discovered that tropical waters contained far fewer plankton than colder, high-latitude waters. This turned out to be due to the behavior of the ocean itself. Colder waters have an almost uniform temperature from top to bottom, whereas in the tropics (and elsewhere in summer) warm surface layers remain where they are and when their nutrients are exhausted by the plankton the tiny organisms die of starvation. In the high latitudes spring and autumn gales disturb the water and bring up nutrients from below so at these times of year there is a constant resupply of food and the plankton thrive and multiply. Wherever upwelling occurs there is a large population of plankton for larger organisms to eat. This accounts for the extraordinarily rich fishing grounds off the coast of Peru, where anchovy feed on plankton and are then in turn eaten by tuna, whose numbers also increase.

The reason for this upwelling of cold, nutrient-rich Peruvian waters is a current about 550 miles wide running north along the Peruvian coast and named after the man who discovered it in 1802: the
117 73 *144* Humboldt[117] Current. The current is caused by a meteorological phenomenon identified in 1835 by the Frenchman Gustave-Gaspard Coriolis, who showed that the Earth's spin deflects objects moving on a north-south trajectory in the northern hemisphere to the right and in the southern hemisphere to the left. This causes the rotational direction of storms to be different in each hemisphere and accounts for the predominantly western direction of winds in the South Pacific. These winds cause the Pacific West Wind Drift in the ocean. When this oceanic movement hits the South American continent most of it passes south of the continental tip but some of it is deflected north by the coastline to generate the Humboldt Current.

It was in 1857 while investigating the pressure gradients in wind that the Dutch meteorologist Christoph Buys Ballot discovered the law named after him: that a person standing in the northern hemisphere with his back to the wind will have the low pressure on his left and high pressure on the right, and vice versa in the southern hemisphere.

Ballot had already made his name with an unusual experiment he had conducted twelve years earlier, on a railroad track near Utrecht. This experiment involved placing a number of horn players alongside the tracks. Ballot took up a position on the footplate of a train, which then moved along the track at forty miles per hour past the players, who all blew the same note. As the train moved Ballot was able to observe that the pitch of the note played on the instruments

appeared to rise as the train approached them and fall as the train
passed and left them behind. This test proved what a professor of
mathematics in Prague had said three years before. His name was
Christian Doppler[118] and his theory was that if either the source of 118 48 *95*
the sound or the observer were moving toward the other the ears of
the observer would receive each sound wave faster. The pitch of the
note would therefore rise because increased-frequency sound waves
produce a higher note. Conversely, as one or other of the subjects re-
treated the sound waves would arrive less frequently and a lower-
pitch note would be heard.

Doppler was primarily interested in this effect (now known as the Doppler Effect) as it appeared to manifest itself in the color of stars. In his paper *On the Colored Light of the Double Stars* he explained the presence of blue and red stars recently observed by astronomers. Because higher-frequency light was bluer, blue stars must be approaching the observer, and since lower-frequency light was redder, red stars must be receding. If the speed of light could be established, the velocity of these red- and blue-light-shifted stars could be worked out. This thought occurred to a French physicist named Armand-Hippolyte-Louis Fizeau (who also worked out the Doppler Effect some six years after Doppler, so the Doppler Effect is sometimes referred to as the Doppler-Fizeau Effect).

Fizeau was an accomplished astronomer, and in 1845 together
with Leon Foucault[119] had taken the first ever daguerreotype pho- 119 40 *75*
tographs of the solar surface. In 1849 Fizeau developed an ingenious
way of calculating the speed of light. He spun a large 720-tooth cog-
wheel on its axis and shone a beam of light between the teeth. The
light beam was reflected by a mirror set some five miles distant. At a
certain speed (12.6 revolutions per second) the speed of the passing
teeth was such that they coincided with the crests in the reflected
light waves and the light disappeared. Working out the math relating
to the frequency of the light, the speed of the wheel, and the distance
to the mirror, Fizeau was able to say that the speed of light was about
196,000 miles per second (only 0.05 percent from its actual speed).

Fizeau married the daughter of Adrien de Jussieu, the last in a line of French botanists. Jussieu succeeded his father as professor of botany at the Natural History Museum in Paris and made a minor contribution to the development of vegetable taxonomy. As a young boy at school in the Lycée Napoleon he struck up a lifelong friendship with Prosper Mérimée, the son of the secretary to the Ecole des Beaux-Arts. Prosper was a sexually precocious young man and was involved in at least one scandal while still at school. His first mistress appears to have been one of his mother's painting pupils, an Englishwoman seven years his senior named Fanny Lagden, who dedicated her life to him and was buried in his grave in Cannes.

In the 1820s, after receiving a law degree, Prosper spent time among the Parisian bohemians, including Stendhal and Alexander von Humboldt, and met the Viollet-le-Duc family, with whose son he would work in later life. Mérimée began his literary work in 1822 with a historical drama about Cromwell. In 1828, while recovering from a gunshot wound received during a pistol duel with an aggrieved husband, he wrote the work that would make his name: *The Chronicle of the Reign of Charles IX,* one of the great French Romantic novels. In 1830 Mérimée visited Spain on a tour of museums and at one point while traveling on a stagecoach fell into conversation with a fellow passenger, the Count de Teba. The count invited Mérimée home to Madrid, and it was there that he met the countess and her little five-year-old daughter Eugenia, with whom he became fast friends. It was reportedly while he was staying with the family that the countess told Mérimée a story about a gypsy girl who stabs her lover in a jealous rage. Later Mérimée turned the tale into a novel that would become famous when its plot was used by•Bizet for his opera *Carmen.*

Returning from Spain, Mérimée became a government bureaucrat and in 1834 was appointed inspector general of historic monuments. It was a job that delighted him. Over the next eighteen years he toured France, restoring some of the country's greatest architectural treasures with the aid of his young friend and architect Eugène-Emmanuel Viollet-le-Duc. In all they restored more than four thousand monuments and buildings, including the great Gothic cathedrals, the Roman theaters of Arles and Orange, the Palace of the Popes at Avignon, the Abbeys of St. Denis and Cheroux and the me-

dieval castles of Blois and Chinon. In 1853 Mérimée's little Spanish friend Eugenia, now grown up (and known in French as Eugénie[120]), **120** 46 *88*
married Napoleon III and became empress of France. She immediately persuaded the emperor to grant Mérimée a life senatorship and an annual pension of thirty thousand francs.

That year, Mérimée mounted a vitriolic attack on the judges who had issued an arrest warrant for Guillaume Libri-Carucci, a friend of Mérimée's and inspector general of French libraries, for stealing rare books. Libri had fled to England taking with him a large number of rare books. Libri was an Italian, and in 1850 Mérimée had gone to London to visit another Italian friend to discuss the possibility of Libri gaining a position on his staff. Mérimée's London friend was named Antonio Panizzi. In 1823 he had arrived in England after fleeing Italy to escape a death sentence for his membership in a revolutionary nationalist group known as *carbonari*. This secret organization, many of whom were imprisoned or executed, was fighting to rid Italy of Austrian occupation, and Panizzi was one of its most senior members. In England he managed to get a job as professor of Italian language and literature at the newly founded University College, London. In addition to this badly paid post he also became assistant keeper of books at the British Museum. His appointment would one day change the life of scholars all over the world.

In the ninety years since its opening in 1753 little had changed at the museum and its services to the public had become woefully inadequate. Most people now regarded the library as a haven for the idle rich. In 1831, when Panizzi was appointed to the Department of Printed Books, it was the least important section of the museum, containing only 240,000 books housed in rooms closed to the public. In 1837 Panizzi was appointed as keeper and immediately began to lobby for more funds. He played on the British sense of national pride, making unfavorable comparisons between the British Museum and other national libraries. The technique worked, but it took until 1846 for action to be taken by Parliament on Panizzi's request for extra money. In the expansion that followed Panizzi found himself with a space problem. He sketched out an idea for the solution: a great circular reading room to be built in the courtyard of the British

Museum, whose walls would contain bookshelves. After the opening in 1857 the reading room became a Mecca for scholars worldwide.

The British Museum had been founded in the first place to house the collections of Sir Hans Sloane, a doctor and antiquarian. After studying in France, where he met the great naturalists, including Pierre Magnol (after whom the magnolia was named), in 1687 Sloane sailed to Jamaica as personal physician to the newly appointed governor Monck. In the two years Sloane spent on the island he avidly studied its flora and fauna, geography, meteorology and local folklore. In the search for new plants for use as drugs or food (he collected over eight hundred), his attention was particularly taken by chocolate. Later he produced the recipe for chocolate mixed with milk, the precursor of the modern chocolate drink. In 1712, after his return to London, Sloane became physician to the king, then president of the Royal College of Physicians, and finally president of the Royal Society following the death of Isaac Newton.

By 1753 when he died his collection of "curiosities" was well-known as "Sir Hans Sloane's Museum." By the standards of the time the collection was very large (twenty-five hundred items, twenty-five thousand coins and medals, and seven rooms full of books) and was regularly visited by leading scholars. In Sloane's will he stipulated that if the nation did not buy the collection from his heirs for twenty thousand pounds within one year of his death it would be offered in turn to the Royal Academies of Science at St. Petersburg, Paris, Berlin and Madrid (each would have a year to decide). After this the collection would be sold to the highest bidder. The British government responded by establishing a lottery to raise the money, and the British Museum opened its doors only six years later. Until 1805 admission was by ticket only and groups were escorted.

Hans Sloane was also an innovative physician and took a leading part in experiments with inoculation for smallpox, a killer disease at the time. In 1716 the Venetian consul at Izmir communicated to Sloane details of a procedure used in Turkey. Sloane did little about it until Lady Mary Wortley Montagu returned to England from a two-year stay in Turkey as wife of the British ambassador. While there she
121 102 *190* had witnessed the inoculation[121] procedure and had became so en-

Fig. 26: *The British Museum, designed in Neoclassical style by Sir Robert Smirke. This present building dates from 1852.*

thusiastic she had her own son inoculated. Part of the reason for her interest was that the disease had already killed her brother and she herself had contracted smallpox in 1715. Thanks to Hans Sloane she had survived, though losing her famous looks in the process.

Lady Mary was the daughter of the Earl of Kingston and had for some years been well-known both for her beauty and her wit. Later in life the poet and satirist Alexander Pope would become infatuated with her, but when the relationship turned sour the two became mortal enemies.

During her stay in Turkey Lady Mary first studied the language and then begun to travel about, disguised in voluminous Turkish costume. Thanks to her sex and rank she was accorded the rare privilege of visiting aristocratic Turkish ladies in their harems, where she learned much about Turkish life and customs. It was during these visits that she came across the practice of inoculation. The Turks would deliberately infect small children with pus from smallpox pustules. Over a few days the children would develop swellings and a number of pustules but then these would heal and the inoculated subjects would never again contract the disease. Lady Mary wrote: "I am patriotic enough to take pains to bring this useful invention into fashion in England; and I should not fail to write to some of our doctors very particularly about it, if I knew any one of them that I thought had virtue enough to destroy such a considerable branch of their revenue for the good of mankind." On her eventual return to England Lady Mary set about persuading all and sundry to take up the matter. Princess Caroline had her two children treated and then everybody else followed suit.

While in Turkey Lady Mary also wrote a collection of essays now known as the *Embassy Letters*. She had apparently never intended them for publication, but after her death a copy of the essays was passed to a printer and the work became public. The letters are delightfully written, full of fresh and vivid descriptions of what Lady Mary saw and thought as she wandered the gardens and palaces of Turkey: "I allow you to laugh at me, for the sensual declaration in saying that I had rather be a rich effendi, with all his ignorance, than Sir Isaac Newton with all his knowledge."

One of her observations related to the profusion of tulips to be found in Turkey. Her visit coincided with an upsurge of tulip mania among the Turks. There were more than thirteen hundred varieties of the flower, with such exotic names as "Beauty's Reward," "Dawn Pink," and "Lover's Dream." The French ambassador wrote back to Louis XV describing the way the palaces were bedecked with the flowers: "The trellises are all decorated with an enormous quantity of flowers of every sort, placed in bottles and lit by an infinite number of glass lamps of different colours. These lamps are also hung on the green branches of shrubs which are specially transplanted for the fête from neighbouring woods and placed behind the trellises. The effect of all these varied colours, and of the lights which are reflected by countless mirrors, is said to be magnificent. The illuminations, and the noisy consort of Turkish musical instruments which accompanies them continue nightly so long as the tulips remain in flower." The tulip had first been introduced to Europe in 1645 when a Flemish scholar, Ogier de Busbecq, brought seeds from Istanbul to Vienna. He also brought the wrong name for the flower. In Turkey the tulip is called *lalé,* but when Busbecq asked what the flower was called he was told it was a "tulipand-flower." *Tulipand* is the Turkish word for "turban," and his informant was describing the shape of the flower. So in the West the name became "tulip."

The Swiss horticulturist Konrad Gesner became the first naturalist to describe and illustrate a tulip in his 1651 *Book of German Gardens.* After spending time teaching Greek at Lausanne Academy, in 1541 Gesner settled in Zurich as professor of natural history and set up a medical practice. Most of his time was spent writing. In 1555 he began work on the two-volume *Opera Botanica,* for which he drew nearly fifteen hundred illustrations. Gesner was the first botanist to recognize the importance of floral structures as an aid to systematic classification. He was also the first to stress the importance of seeds, showing that they often revealed connections between apparently unrelated plants.

Gesner was also interested in bibliography, and in 1555 published the three volumes of his *Universal Library,* a monumental work that contained a list of all books published since the invention of the printing press a hundred years earlier, as well as a catalogue of authors arranged alphabetically together with brief descriptions of their works, and a giant dictionary divided into twenty-one subjects (including grammar and philology, dialectics, medicine, astrology, geography and theology). The third volume included an account of the 130 known languages and translations of the Lord's Prayer in twenty-two of them. Gesner also wrote on the importance of the use of textual analysis to understand ancient texts such as the Bible. This endeared him to his godfather, Ulrich Zwingli, the man who led Switzerland to Protestantism.

Zwingli had an uneventful early life, matriculating at the University of Vienna and becoming ordained as a priest in Constance, Switzerland, in 1506. He then spent ten years as a serving cleric in the small Swiss town of Glarus. There in 1515 he met the man who was to change his life: the great Dutch humanist Desiderius Erasmus. Erasmus introduced Zwingli to the historical, analytical approach to the study of biblical texts. Zwingli began to take an increasingly critical attitude to the way the Catholic Church conducted itself and preached the faith, and his fame as a scholar began to spread. On January 1, 1519 (his birthday), partly in recognition of his publications, Zwingli was given the post of "people's priest" at the Great Minster in Zurich and found himself in a position of power. In a city of only six thousand adults the pulpit was stage, loudspeaker, radio, newspaper, television and Internet combined. One eminent church administrator was said to have advised his supporters that if they wanted to influence policy they had to be sure to get their proposals accepted "before the preacher stands up in the pulpit."

Zwingli's break with the Catholic authorities came over the next four years with his attacks on belief in purgatory, the invocation of saints, monasticism, indulgences, tithes, ecclesiastical vestments, the Mass, Latin services, music, baptism, transubstantiation and celibacy. His most public act of defiance, other than getting married in 1522, took place on the evening of the first fasting Sunday in Lent, March 9 the same year. That evening a group of Zurich citizens disregarded Catholic teaching that forbade the eating of meat during the festival and ate smoked sausages. This event took place at dinner in a private house at which Zwingli was present, although he himself did not eat the offending sausages. Two weeks later his sermon was titled: "Regarding the Choice and Freedom of Foods." In it Zwingli cited the Bible as supporting the view that Christians were free to eat all foods since these were in themselves neither good nor bad. In this way Zwingli reduced the act of fasting to a matter of private conscience, expressing the humanist belief that matters of faith should be left to the individual, who would be guided by revealed truth. In 1525 the Zurich Town Council introduced new, stringent "Zwinglian" laws against prostitution and new regulations for social behavior and dress. It became a civil offense to blaspheme, play card games or dice or to wear silk, gold, silver, velvet and low-cut shoes. In 1530 a general curfew closed all inns at 9:00 P.M.

Zwingli brought about one other major change to life in Zurich. For centuries it had been a Swiss custom to send young men off to fight as mercenary soldiers. Fighting was popular and admired in Switzerland, where life at home was hard and monotonous and where the lack of cultivable land offered little opportunity for younger sons. Swiss mercenaries had long been held in high esteem throughout Europe. They had developed a particular fighting style, thanks to their use of pikes. A Swiss pike "square" of several thousand men massed together and moving as one, ready by sheer weight of numbers to roll over an enemy formation, or to stop suddenly and angle their pikes outward in defense, was well-nigh invincible. Enemy cavalry could do little, since before they were able to cut down a pikeman they could be impaled on his pike. Zurich had a long-term agreement to supply mercenaries to France, and Zwingli persuaded the Canton Council to revoke it.

The pike square was already on the way to obsolescence. A few years before, at the Battle of Marignano (in which Swiss mercenaries had fought for the French against the Spanish) a new firearm, the arquebus, had radically changed the nature of war and rendered the pike square ineffectual. Toward the end of the seventeenth century other new developments in weaponry accelerated this process.

The first was the flintlock musket. It was designed so that pressure on the trigger released a sprung arm carrying a flint, which then struck a small, serrated metal plate. This caused the flint to shoot

Fig. 27: *Swiss mercenary pikemen (wearing their white cross) fighting for the French at the 1525 Battle of Pavia, Italy.*

sparks into a pan filled with powder at the same time as the action also opened the pan cover. When the powder in the pan ignited it lit the full charge of powder inside the barrel and fired the ball.

The slightly later development of a prefilled paper cartridge (the musketeer opened it with his teeth, poured the powder into the barrel and then inserted the ball on top of it) sped up the rate of fire to two or three rounds per minute. This was twice as fast as the earlier matchlock musket, in which the powder was ignited with a smoldering fuse. With no further danger of accidental discharge from sparks (which could happen when matchlock musketeers stood close to each other), the new flintlock permitted soldiers to stand less far apart. This in turn led them to discard the old broad-brimmed hats and full-skirted coats in favor of slim-fitting clothing. The flintlock enabled close ranks, firing by rows, to maintain a hail of fire.

The second innovation was the new bayonet, which consisted of a blade attached to a metal sleeve that fitted around the muzzle leaving the musket free to fire with the bayonet still attached. The infantryman was now pikeman and musketeer in one. He could shoot at the pike square from a distance and then move in with the bayonet to finish the job.

These innovations combined to reduce the need for Swiss mercenaries. The French minister for war, the Marquis de Louvois, realized that the new weapons and the tactics involved in their battlefield use were going to require soldiers to be better-trained than ever before. Such training was going to take several years, so the new army was going to have to be professional and permanent.

The idea of a standing army had already been developed to some
122 63 *129* extent in England, Holland[122] and Sweden, but Louvois made his force fully national and introduced organizational measures that made the French army perhaps the most advanced on the Continent. Louvois created a quartermaster's department with a commissariat staff to supervise the price, quality and transport of supplies. This in turn led to improvement in the roads over which the army traveled and to the establishment of strategically placed magazines for arms, ammunition and food. Louvois also found a man who was so good at drilling the troops his name became a household word: Jean Mar-

tinet. Other changes included the issue of uniforms, the rationalization of medals and awards, an organized promotion system, regular pay, codes of discipline and the construction of the Hôtel des Invalides in Paris for those crippled in action.

Toward the end of the century an Italian immigrant to France added one final element to the military mix by introducing the first marching music. In 1661 Jean-Baptiste Lully was appointed superintendent of court music to Louis XIV and began to write music for dances in which the king would perform starring roles. In 1672 Louis added a school of dance to his Royal Academy of Music. One year before, Pierre Beauchamp had been made royal dance master, and it was he who facilitated the development of ballet with his new system for learning steps called "choreography."

By the time Beauchamp retired in 1687 he had laid the foundations of *danse noble* and made the French style of dance and the terminology it employed standard until modern times. In 1701 a book by Raoul Feuillet replicated much of Beauchamp's work. Titled *The Art of Describing Dance,* it described the "track notation" of Beauchamp's choreographic technique, which used a line to indicate the path to be followed by the dancer. Black dots to right and left of the line indicated the positions of the feet. Additional strokes and symbols indicated arm movements and any extra ornamentation to be added. The new choreography would be in use for nearly a hundred years. By 1706 Feuillet had been translated into English by a dance master named John Weaver.

Weaver had already established himself in London as a theatrical dancer in pieces performed supplementary to the action of a play. In 1702 he had produced his *The Tavern Bilkers* at Drury Lane Theater. This work was described by a contemporary as "the first entertainment that appeared on the English stage where the representation and story was carried on by dancing, action and motion only." In 1717, Drury Lane also saw the production of more of Weaver's danced mimes: *The Loves of Mars and Venus* and *Harlequin Turned Judge,* among others. Drury Lane Theater was at the time in fierce competition with the newly redecorated Lincoln's Inn Fields Theater managed by John Rich, who experimented with pantomime and dance. In 1728 Rich staged a new production by the relatively unknown author John Gay, titled *The Beggar's Opera.* It combined dance, popular song, comedy and political satire, and the story line contained a thinly veiled attack on Prime Minister Robert Walpole. On January 28, 1720, *The Beggar's Opera* attracted an opening night crowd of twelve hundred. Before the season was over the ballad-opera had been performed an unheard-of sixty-two times and was a smash success. It was said at the time that the satirical *Beggar's Opera* "made Rich gay and Gay rich."

Satire was a popular genre in a time of transition from rule by monarch to rule by Parliament, when corruption was rife. This was
123 95 *177* the heyday of Alexander Pope[123] and Jonathan Swift,[124] two of the
124 90 *162* greatest masters of satire in English literature. Gay knew both of them and for a brief time in 1714 they were all fellow members of a private writers' association named the Scriblerus Club. Scriblerus members met informally at each other's lodgings to eat, drink and regale the assembled company with pieces satirizing the great and the powerful. The official purpose of the club was to write the memoirs of a fictitious individual named Martin Scriblerus, but the real intent was to publish pieces under his name attacking the follies of the self-styled intellectuals of the day. Targets abounded in a period when people still believed in witchcraft, the philosopher's stone and astrology.

The club member who did most of the wining, dining and entertaining (because as the royal physician he could afford to do so in his grace-and-favor apartments at St. James's Palace) was Dr. John Arbuthnot, whose other obsession, besides satire, was statistics. In 1692 Arbuthnot published *On the Laws of Chance,* in which he made his famous remark: "There are very few things which we know, which are not capable of being reduc'd to a Mathematical Reasoning; and when they cannot, it is a sign that Knowledge of them is very small and confus'd." In 1710 he wrote *An Argument for Divine Providence,* in which he attempted to demonstrate that providence and not chance governed the sex of a child at birth. From a study of the mortality tables (births and deaths) for London over a period of eighty-two years he calculated that there were many more births of boys than girls. Since this was against the laws of chance Arbuthnot suggested that because men lived riskier lives than women the higher male numbers showed God was making sure enough men would survive to renew the population.

This first known example of statistical inference attracted the attention of continental scientists and in particular that of a young Dutch mathematician, Willem 'sGravesande. 'sGravesande met Arbuthnot during a one-year visit to England in 1715, when he also met (and impressed) Isaac Newton, whose work he was already teaching in Holland. Newton was evidently not the only one to be impressed by 'sGravesande's work. At one point the great French thinker Voltaire made a special trip to visit him in Holland to get his opinion on the book Voltaire was writing about Newton.

At the time, Voltaire was in residence with the love of his life, the beautiful and brilliant Marquise Emilie de Châtelet, at the Château de Cirey in Eastern France. The house was a long way from Paris, where over the previous fifteen years Voltaire had established his reputation as France's greatest writer and found himself in serious trouble with the authorities for his outspoken views. There had been particularly negative reaction to his *Letters Concerning the English Nation,* written in 1729 after a three-year visit to London (where he had attended the funeral of Isaac Newton). The book praised the liberty accorded writers in England and made unfavorable comparisons with the absolutist regime in France.

In 1733, after numerous love affairs, Voltaire met Emilie. His memoirs begin with the sentence: "I was weary of the idle and turbulent life of Paris, of the crowd of fops, of the bad books printed with official approval and royal privilege, of literary cabals, of the meanness and rascality of the wretches who dishonoured literature. I found in 1733 a young lady who felt more or less as I did, and who resolved to spend several years in the country to cultivate her mind, far from the tumult of the world." Voltaire went with Emilie to Cirey and they set up house together. Emilie's husband was an army officer who was almost permanently absent and who seems to have approved of his wife's liaison. Each of these two workaholics had a separate study, and Voltaire also kept a small lab. Both were writing about Newton's work. Voltaire was producing a popular version and Emilie was tackling the mathematics. In spite of the fact that Voltaire and Emilie lived informally and did not believe in servants, the cream of the European intelligentsia was soon beating a path to their door. Staying at Cirey was not a relaxing experience. At all hours of the night and day guests were expected to take part in magic lantern shows, philosophical discussions, plays and poetry readings. In 1738 Voltaire's *Elements of the Philosophy of Newton* went into print and was an immediate and sensational success. In London it gained Voltaire election to the Royal Society. In 1758, a few years after Emilie had died, Voltaire bought a villa at Ferney in Switzerland and retired "to cultivate his garden." It was there in 1765 that he received an intriguing letter from an Italian scholar about the souls of snails and worms.

The writer, Lazzaro Spallanzani, was professor of natural history at the northern Italian University of Pavia where he had for some years been investigating the regenerative properties of certain animals: snails, worms and salamanders. He had discovered he could cut pieces off these creatures and the missing parts would grow back. In the case of some worms this might sometimes result in the regeneration of two separate creatures. This created a theological problem, because if souls were indivisible then when two worms were created out of one and each of them had a soul, where did the extra soul come from? Spallanzani said it must always have been present in an egg of some kind. The remark started the whole of modern reproductive physiology.

The other work in which Spallanzani made startling advances related to the contemporary view, held primarily by the English microscopist John Needham, that there was a "vegetative force" present in the minute organisms that could be seen down a microscope. This, Needham claimed, would account for the presence of maggots in cheese, moths in carpets and so forth. According to Needham such organisms were created spontaneously by the vegetative force in the cheese and the carpets. Spallanzani disagreed and set about proving Needham wrong. In 1761 he took a flask of dirty water in which he had previously seen microscopically small organisms, boiled the

water in the flask and then melted the neck of the flask to seal it. Some time later he broke the neck and quickly examined the water under his microscope. The organisms in it were dead. But even as he watched live ones appeared. Since these organisms only appeared after the flask was opened, Spallanzani concluded they had to have entered the flask from the air. This was almost exactly the same experiment for which Pasteur[125] would get all the credit a hundred **125** 54 *114*
years later. **125** 3 27

Spallanzani's reputation was so great that he inspired the character of the "wizard-scientist" in a German story written by E. T. A. Hoffmann and immortalized in Delibes's ballet *Coppélia*. Hoffmann began as a lawyer and became in turn a theater manager, then a writer of ballets, operas and novels. Hoffmann's novels were among the first "psychological" stories, grotesque tales full of doppelgangers and psychopaths. Some of the plots went into musical form in Wagner's *Die Meistersinger*, Offenbach's *Tales of Hoffmann* and Tchaikovsky's *The Nutcracker Suite*.

In 1816 Hoffmann was appointed counselor to the Berlin Court of Appeal. Two years later he was in charge of an investigation regarding the activity of Friedrich Jahn, who was accused of secret and treasonable association with intent to overthrow the government and subsequently jailed for six years. Jahn was a nationalist who had reacted to Germany's defeat at the hands of the French in 1806 by starting a movement to persuade Germans from every principality and state to unite and establish a single, liberal nation. Jahn's technique for rallying supporters was to set up gymnastic clubs with the aim of inculcating youths with the sense of discipline, comradeship and obedience that would be required for the wars that undoubtedly lay ahead as Germany asserted herself. Jahn's clubs sprang up all over Germany and soon became hotbeds of subversion and free speech.

In 1819 the political murder of a prominent German conservative, August von Kotzebue, brought a crackdown by the Prussian authorities that closed the gym clubs and suppressed free speech. One of Jahn's followers, Adolf Follen, was arrested and tried for distributing subversive literature. Although Adolf would eventually be acquitted, his brother and fellow liberal Karl decided in 1820 to flee the coun-

Fig. 28: *An early nineteenth-century German gymnastic club. The athletes usually dressed in loose cotton jackets and trousers.*

try, first to Switzerland and then in 1824 to America. In 1825 he settled in Cambridge, Massachusetts. As soon as he arrived he began gymnastic exercises with the students at Harvard, and in one of the university dining halls opened the first college gym in America. At the same time two more gyms opened nearby, one in Boston and the other in Northampton, both run by fellow Germans. By 1850 there were a hundred gym clubs in America run by German refugee immigrants, most of them liberals or socialists.

One of the first American organizations to adopt gymnastics as part of its program was the YMCA, the first of whose branches opened in Boston in 1851. That year the YMCA received a letter from a man in Geneva who was the corresponding secretary for a group of young Swiss Christians, suggesting that such groups form an international association. In 1855, partly at his instigation, the first World Conference of YMCAs occurred in Paris, attended by representatives of Belgium, France, Britain, Canada, Germany, Holland, Switzerland and the United States. The conference established the World Alliance of YMCAs. Henri Dunant, the Genevan who had suggested the conference, was the main author of its charter.

126 26 *50* Four years later Dunant found himself in the small northern Ital-
126 116 *218* ian village of Solferino for a meeting with the French emperor
126 129 *246* Napoleon III,[126] who was about to watch a battle between his army

and that of the Austrians. Dunant stayed to watch, too. On June 14, 1859, 350,000 soldiers from both sides met and proceeded to slaughter each other. Over forty thousand men were either killed or wounded. Dunant, watching from the hill, was appalled by the savagery. When the battle was over, in the nearby town of Castiglione delle Stiviere, Dunant organized the townsfolk to help the wounded of both sides. For three days and nights Dunant and his helpers tried desperately to save hundreds of young men from death.

Three years later Dunant published *A Note on Solferino,* in which he wrote: "It ought to be possible in peacetime to get together trained helpers who would care for the wounded after a battle. People who would be ready to go and help wherever and whenever they were needed. . . . Countries at war should recognise these helpers and give them all possible assistance. . . . A meeting must be held where these ideas could be discussed." After intense lobbying of the kings, queens, generals and prime ministers of various European countries, in 1864 Dunant succeeded in arranging a meeting in Geneva, where the Red Cross was formed and the Geneva Convention on the treatment of wounded soldiers was signed.

For the Red Cross one of the earliest and most urgent operations required on the battlefield was the blood transfusion. This generally involved the difficult process of joining donor and recipient blood vessels and usually ended in the inexplicable death of the recipient. The mystery was solved in 1900 by an Austrian physician named Karl Landsteiner, when he discovered that blood from one person could cause another's red blood cells to "clump." Clumping could cause blockages in the capillary system, bringing damage and even death. Landsteiner found that blood contained two "factors" that could cause clumping. He named the blood groups containing the factors "A" and " B." People possessed either one, or both, or neither. So a person's blood group was either A, AB, B, or O. In 1900 Landsteiner mentioned this idea in a footnote and won the Nobel Prize.

The Nobel Prize would also go to the man who solved the other half of the transfusion problem: the difficulty in uniting blood vessels. He did so with a technique that was astonishing in its simplicity. He used three sutures to join the blood vessels at three equilateral points around the vessel's circumference. Then he pulled on two of the sutures, causing the edge of the vessel between them to straighten, and sewed the straight edges together. He then repeated the action twice more and released the vessel to spring back circular, and joined. The name of the medical stitcher was Alexis Carrel, and like Landsteiner he spent time at the Rockefeller Institute for Medical Research in New York. Carrel's ultimate aim, in finding a way of suturing blood vessels, was to be able to carry out organ transplants. For this he also needed to be able to keep a separated organ temporarily supplied with oxygen and nutrients.

The man who made this possible in 1930 worked for several years with Carrel before perfecting a sterilizable glass perfusion pump with which Carrel was able to keep a kidney alive for several weeks. The pump-maker was Charles Lindbergh, who three years earlier had completed the first ever transatlantic crossing in his monoplane

Fig. 29: *Charles Lindbergh, standing before his monoplane* Spirit of St. Louis, *in May 1927, just after his epic flight.*

Spirit of St. Louis. Lindbergh married the daughter of U.S. Ambassador Dwight Morrow. After the wedding Morrow went to London to attend a disarmament conference. One of the things the conference did was to reaffirm the constraints imposed on the German navy by the 1919 Versailles treaty. Germany was limited to building three new ships, all of them under ten thousand tons' displacement.

The first, *Admiral Graf Spee,* was launched in 1936. *Graf Spee* circumvented the intentions of the disarmament agreement because she and her two sister ships packed the punch of a full-size battleship and went faster and farther than a cruiser. Scarcely had World War II begun before *Graf Spee* sank nine British ships and took their crews on board her supply vessel *Altmark. Graf Spee* was finally run down by the British in December 1939. After a brief firefight the German battleship took refuge in neutral waters off Montevideo harbor, Uruguay. The Uruguayans had given the ship four days to leave, and the German High Command ordered the captain to scuttle her. The British turned their attention to *Altmark.* Two months later she was found in a Norwegian fjord and the British prisoners were rescued.

Hitler took this invasion of neutral territory as an indication that the British were about to invade Norway, so he advanced his own invasion plans. On April 8, 1940, the Germans invaded Norway. Three weeks later they had placed an exceptionally tight security ring around the hydroelectric power station at Vermork in the Rjukan valley to the east of Oslo. They had done so because hydroelectric power was essential to a top-secret Nazi research project that involved a special kind of water.

One essential factor in causing a nuclear chain reaction is to be able to slow down the passage of a neutron so that it does not pass through the nucleus of an atom of uranium in a trillionth of a second. At a slower rate there is more chance that it will collide with the nucleus of a uranium atom and split it, releasing particles that will in turn split other nuclei and so on. When uncontrolled, this "chain reaction" process is what causes a nuclear explosion.

The material which can slow the initial neutron bombardment was produced at Vermork. It was called "heavy water," because with the use of a gigantic amount of electricity water could be "reduced." This would leave it with a heavier-than-usual amount of deuterium, an element occurring naturally in water. If the heavy water were placed between the source of the neutrons and the uranium nuclei, the deuterium atoms would act to slow the neutrons enough to trigger a chain reaction.

This was why on the night of February 27, 1943, a group of Allied commandos (all of them Free Norwegians) entered the Vermork plant and blew it up. When they hit the water, Hitler was denied the material that might have given him an atomic bomb.

CHAPTER 10

In Touch

On March 25, 1951, *The New York Times* ran a front-page report carrying the astonishing news that Argentina had successfully operated a nuclear fusion reactor. It appeared that in the late 1940s the Argentine dictator Juan Peron, increasingly estranged from his own scientific community, had set up an isolated island laboratory for Ronald Richter, a German scientist. In this laboratory on February 16, 1951, according to an Argentine press release, "There was held with complete success the first tests which, with the use of this new method, produced controlled liberation of atomic energy." No further details were available.

The reaction of European and American researchers was predictably skeptical. The harnessing of fusion power presented extraordinarily technical problems, only one of which was to reproduce the conditions on the surface of the sun where at temperatures in excess of two hundred million degrees Fahrenheit the process of fusion converts mass into energy at the rate of five million tons a second. In the solar environment lightweight hydrogen nuclei are so hot and moving at such high rates that when two of them collide they fuse. The union forms a nucleus of heavier helium, and as it does so releases large amounts of energy in the form of light, heat and neutrons. The sun's gravitational field is so massive (three hundred thousand times greater than Earth's) that it holds together with a density ten times that of lead the incandescent gas in which the reactions occur. So the hydrogen nuclei are packed together long enough to ensure that collisions happen with great frequency. Fusion therefore requires three conditions: long confinement time, high density and heat.

These were some of the considerations in the mind of an American scientist named Lyman Spitzer, Jr., when he heard of the Argentine announcement during a skiing holiday in Colorado. During the lengthy intervals spent on the ski-lift Spitzer was stimulated enough by the news to give further thought to fusion, because as an astrophysicist he was acquainted with the stellar fusion process and had earlier become involved in theoretical work on the project to develop a hydrogen bomb.

Spitzer knew how it might be possible to heat a gas to the necessary temperature because he had recently read work by the Swedish physicist Hannes Alfven regarding the influence of magnetic fields on hot gases in the cosmos. A superheated gas would become electrically charged ("ionized"), and its ionized particles would then be attracted to magnetic fields. Spitzer also knew that no material on earth could withstand two hundred million degrees Fahrenheit. The theoretical solution to the problem was to contain the hot, charged gas (known as a "plasma") in a magnetic container. A month later Spitzer presented his ideas for an experimental fusion generator to the U.S. Atomic Energy Commission in Washington, D.C. The device, called a "stellarator," was designed to heat the plasma by inducing a current into it and containing the superhot plasma in a closed, figure-eight-shaped tube inside which a complex grid of magnetic fields would be generated. Spitzer's stellarator was the first of several experimental fusion reactors that would be developed from then till the end of the twentieth century.

The value of developing a successful fusion reactor some time in the twenty-first century is hard to exaggerate. The high-energy neutrons released by the fusion process could be used to generate heat to boil water whose steam would drive electric power turbines. Today the most recently developed fuels for the fusion process, deuterium and tritium, are found in abundance. A fusion reactor is safe because in the case of catastrophic failure the magnetic confinement environment collapses and the fusion process stops immediately. Fusion power plants would also produce no pollutants and could substantially reduce the amount of fossil fuels at present being burned in conventional power plants. There would also be little or no problem with nuclear waste, since fusion produces very much less radioactive by-product than the atomic fission process.

Superconductivity could make fusion even more attractive. Superconducting materials might permit the transmission of electricity at virtually no cost, since cables made of superconducting materials would offer millions of times less resistance to the passage of electric current than conventional cables. In terms of power transmission this would mean that over long distances no intermediate booster power stations would be needed.

Superconductivity was first discovered in 1911 by a Nobel Laureate, the Dutch professor of experimental physics Heike Kammerlingh Onnes, while working at Leiden University in Holland. Onnes was obsessed with extreme cold, and the cryogenics work of his lab was soon to lead the world. Once helium gas had been liquefied by James Dewar in 1898 Onnes decided to use it in the investigation of what he thought might happen when the temperature of certain materials reached close to absolute zero, or minus 459 degrees Fahrenheit. It had been theorized by Walter Nernst in Berlin that as a pure metal became colder its resistance to electricity should become smaller and smaller until at absolute zero it disappeared completely. After early experiments with platinum and gold Onnes became aware that the slightest trace of impurities in the material would reduce the extent to which resistance fell. Onnes realized that the best metal to use would be mercury, since it is liquid at room temperature and is easy to distill and redistill until an extreme degree of purity is reached.

In 1911 Onnes discovered that the resistance of mercury dropped sharply just above the boiling point of helium (and well above absolute zero). Just below this temperature the mercury resistance disappeared totally. Experiments with a superconductive coil of lead immersed in liquid helium produced an extraordinary result. Once a current had been set up in the lead ring, even after the power had been switched off, so long as the lead ring was kept at superconducting temperature the current remained in the lead. Onnes called this a "persistent current" and went on to maintain such a current for a period of two years until he discontinued the experiment.

The liquefaction of gas, so helpful to Onnes, had first been done thirty-four years earlier by a Swiss researcher named Raoul-Pierre Pictet and simultaneously by a Frenchman named Louis Paul Cailletet, whose contribution to cryogenics was triggered by an accident. Cailletet became involved with gases while running his father's blast furnaces, when he became interested in ways of retrieving the materials in the fumes given off by the smelting process. In 1877 he began work on liquefaction. At the time there were six gases considered to be permanent (gaseous in their natural state): oxygen, nitrogen, hy-
127 39 *74* drogen, acetylene,[127] nitrogen dioxide and carbon monoxide.

127 57 *119* Cailletet began with acetylene, and it was at this point that the accident happened. He had theorized that a pressure of sixty atmospheres would liquefy the gas but before this pressure was reached his apparatus sprang a leak and the pressure on the compressed gas suddenly fell. Cailletet had been watching the glass cylinder in which he was compressing the gas and noticed that at the moment the pressure had suddenly fallen a faint mist had appeared. He immediately realized that the fall in pressure was causing the condensation of the gas and producing small liquid droplets. Armed with this knowledge he now succeeded in recreating the same pressure-drop conditions with the atmospheric gases, beginning with oxygen. On December 2, 1877, he compressed oxygen to about three hundred atmospheres, first reducing its temperature to minus 27 degrees Centigrade by surrounding it with evaporated sulphur dioxide. When Cailletet released the pressure in the same sudden manner as had happened in the acetylene experiment, the oxygen condensed and formed liquid droplets.

In 1889 Cailletet's interest in other aspects of pressure was unex-
128 79 *150* pectedly served by the construction of the Eiffel Tower[128] on which he subsequently installed a nine-hundred-foot manometer, running up the tower. The manometer consisted of a transparent tube filled with various liquids. The lower end of the tube was connected to a

pressure source and the upper end left open to the atmosphere. Cailletet was able to calculate the total pressure exerted by each of the liquids tested.

The tower made possible other air-pressure-related experiments when its designer, Gustav Eiffel, dropped small, variously shaped plane surfaces with fine wires attached to them to measure the speed of their fall and experimentally confirmed that air resistance increased as the square of the surface of the object moving through it. In 1906 Eiffel built a wind tunnel at the foot of the tower and for the first time proved that more lift was generated by air flowing over the camber of a wing than striking its underside.

Eiffel knew a great deal about the movement of air because by the time he built the tower he was France's greatest engineer, specializing in high-level railway bridges over gorges and rivers in France, Portugal and Indochina. Eiffel's bridges were miracles of delicate iron tracery, capable of withstanding high wind-loads. By 1886, when the French government decided to build the tallest tower in the world as a centerpiece for the 1889 Paris Exposition, Eiffel was the only engineer with sufficient expertise in wrought-iron to do the job. Given the likely wind-load stresses on the tower cast-iron would have been too brittle and steel so flexible it would have produced an intolerable amount of sway.

Eiffel's specialty was latticework, and the Eiffel Tower is a beautiful example. Eiffel reduced the metal structure to the absolute minimum consistent with safety. He also achieved absolute precision (rivet holes, for instance, had to be placed within one-tenth of a millimeter) by placing the tower foundations on hydraulic jacks that would raise or lower each of the tower's sixteen columns accurately enough to be sure that the piers were absolutely horizontal. This ensured that by the time the structure had reached its full height of nearly one thousand feet it was still exactly vertical.

It was Eiffel's wind-load expertise that had won him a rather special commission just before he got the contract for the tower. The project was nothing less than an attempt by the French government to bring political stability to the country through the ceremonial donation of an extraordinary gift to the United States.

The plan originated with Frédéric Bartholdi, who in 1871 was sent to the United States as the agent of a small but influential group of moderate Republican intellectuals who were concerned that France was in danger of social upheaval following the recent defeat at the hands of the Germans. This had led to the flight of the emperor
129 26 *50* Napoleon III[129] and the inauguration of the Third Republic. French
129 116 *218* democracy was on shaky ground, with monarchists calling for a re-
129 126 *236* turn to the days of the empire, revolutionaries who wanted an extreme left-wing state and the moderates caught between the two. Bartholdi's plan was to rally French public opinion to the moderates with a very public act linking the fledgling French republic with the great transatlantic democracy whose independence the French had virtually assured a hundred years before with troops and money.

The political link with America would be forged by ensuring that contributions from both countries would pay for the construction of France's gift: the Statue of Liberty. And as the statue was to stand exposed to fierce gusts in New York Harbor the ideal person to build the statue's support structure would be Gustav Eiffel. When the Statue of Liberty was finally dedicated on October 28, 1886, the American cofunding was almost entirely due to last-minute efforts by Joseph Pulitzer, owner of the *New York World*. Pulitzer lobbied strenuously to gain public support for the project, which many saw as an irrelevance, by printing in his newspaper the names of every donor, no matter how small the amount given. This would later generate the myth that the statue had been paid for by schoolchildren.

The French intended the statue to be described as "Liberty Enlightening the World," the idea being that it would act as a powerful and very visible reminder to all Americans of the quality of French culture and of America's political debt to France. At one stroke the entire political meaning of the statue was changed by a young Jewish poet named Emma Lazarus. In 1883 well-known authors were invited to write verses about the statue and to permit their poems to be auctioned. Lazarus wrote the poem, which was then chosen for recitation at the dedication ceremony and later inscribed on a plaque placed on the statue's pedestal, the section of the monument built and paid for in America. The last lines of Lazarus's poem turned the

Fig. 30: *Emma Lazarus. Her poem about the Statue of Liberty was titled "The New Colossus."*

Statue of Liberty from a French icon into the symbol of America as the home of freedom:

> "Keep, ancient lands, your storied pomp!" cries she
> With silent lips. "Give me your tired, your poor,
> Your huddl'd masses yearning to breathe free,
> The wretched refuse of your teeming shore,
> Send these, the homeless, tempest-tost to me.
> I lift my lamp beside the golden door!"

Lazarus did one other thing that placed her in the history books. Almost singlehandedly she started the Zionist movement to establish a Jewish homeland in Palestine. Her efforts were triggered by the horrifying news of the 1881 anti-Semitic pogroms in Russia and Germany, when thousands of Jews died in violence during which their homes were destroyed and their possessions confiscated. Forty thousand of the survivors emigrated to the United States, where Lazarus spoke out strongly in newspaper articles and poems against the persecution the new Jewish immigrants had already suffered and against the conditions in which they were being held on Ward's Island before entry to America.

In 1882 she wrote: The Jews . . . "must establish an independent nationality." That year she heard from an Englishman who had for three years been taking practical steps in the same direction. Laurence Oliphant, an English non-Jew, wrote to Lazarus from Palestine, where he had been trying to obtain suitable land and the permission of the Turkish occupying authority to build settlements for European Jewish refugees. His letter to Lazarus was an appeal for her help in getting the U.S. government to ask the Russians to persuade the Turks to allow Romanian Jews to settle in Palestine.

130 34 *67* In 1888 Oliphant married Rosamund Dale Owen,[130] niece of the man who had brought the Smithsonian Bill to the floor of the U.S. Senate. Before the couple could retire to their new home in Palestine Oliphant died. He had had an extraordinary life. Beginning as a lawyer in Sri Lanka, where his father was the British chief justice, in 1853 Oliphant had taken up travel writing and was hired by the *London Daily News* to cover the runup to the Crimean War. In 1854 Oliphant traveled in Canada and the United States. The following year he reported on the siege of Sebastopol for *The Times* of London. Another trip to the States followed. In the decades that followed Oliphant visited China, Japan, Korea, Italy, Poland, Moldavia, Albania, France, Germany and finally Palestine.

In 1857 while in China he served as private secretary to the eighth Earl of Elgin at the time when Elgin was using gunboat diplomacy to force the Chinese to accede to British demands that they open up China to foreign trade and accept the legalization of opium (which the British would then import to China from India, as payment for goods). The Chinese agreed, then reneged on the deal, so Elgin was obliged to return to China and force the imperial government into submission by bombarding the Summer Palace in Peking. Elgin found the violence, especially the damage to the ancient palace, extremely regrettable.

Elgin's father had held similar views about historical monuments. In 1799 the seventh earl was appointed ambassador to Turkey. In keeping with the new craze for things ancient and Greek (and with
the help of Sir William Hamilton,[131] antiquarian and husband of **131** 13 *37*
Emma, who ran away with Lord Nelson), Elgin obtained the permission of the occupying Turkish authorities to erect scaffolding around the Athenian Parthenon. The purpose of the scaffolding was to make plaster casts of various carvings. Elgin was also given permission "to take away any pieces of stone with old inscriptions or figures thereon." The Parthenon, on the Athenian Acropolis, was a Doric temple built between 447 and 432 B.C.E. and was the crowning glory of Pericles' program of public works designed to establish Athens as an imperial city at the head of a confederacy of city states.

By the beginning of the nineteenth century the glory that was Greece had all but vanished from Athens, by now a dirty, provincial slum city of perhaps twelve hundred houses. The Parthenon itself was in ruins. It had been turned into a mosque in the fifteenth century, then became a gunpowder magazine subsequently wrecked by a lightning strike that exploded its contents. In 1687 Venetian artillery had blown off the temple roof and destroyed parts of the colonnade. By 1800 the Turks were carrying off large pieces of the temple and statuary to grind up for lime to make mortar. When Elgin saw the magnificent (and still relatively undamaged) frieze running around the entire temple inside the colonnade and the untouched metopes (four-foot panels, carved in high relief) on the exterior he decided not to make plaster casts of them but to take away both metopes and frieze for safekeeping.

The removal of what would become known as the "Elgin Marbles" took more than nine years at a cost that almost bankrupted Elgin. His actions did not meet with universal approval in England. In
Childe Harolde Byron[132] described Elgin as "the last, the worst, dull **132** 30 *61*
spoiler" taking "the last poor plunder from a bleeding land," and in **132** 60 *122*
1815 he wrote an entire poem about Elgin's stripping of the Parthenon, describing Elgin as "cold as the crags upon his native

coast. His mind as barren and his heart as hard." However, there was little argument that the marbles were, in the words of the evidence presented before the House of Commons Select Committee set up in 1816, "the finest things ever to come to this country." In the end it was decided that the government would buy the marbles from Elgin for the sum of thirty thousand pounds and place them in the British Museum. Elgin was in no position to refuse, but he had spent an estimated seventy-four thousand pounds on removing, transporting and storing the marbles, and the cost of the entire venture would financially ruin his family for the next two generations.

One of the experts called before the 1816 Select Committee had been Sir Thomas Lawrence, who argued in support of the decision to purchase. By this time Lawrence was the country's most famous portrait painter, his recent sitters including the Prince Regent and the Duke of Wellington. Lawrence had begun as a child prodigy, earning large fees for portraiture by the age of eleven. In 1787 he was enrolled at the Royal Academy and began to paint the rich and famous.

Fig. 31: *Lawrence painted Queen Caroline, who was an accomplished sculptor, with her chisel in hand.*

Lawrence rapidly became known for his inability to treat his subjects with the fawning respect to which they were accustomed. On one occasion when his subject, the tsar of Russia, complained how unreasonably long the work was taking Lawrence replied: "Sir, I can't be reasonable!" In 1789 Lawrence was invited to paint Queen Caroline and Princess Amelia, whose portraits were then exhibited to great acclaim in the Royal Academy exhibition of 1790. Two years later Lawrence was appointed painter to King George III.

This was at a time when the King's health was failing. His surgeon was a Scotsman, John Hunter, who had reached the age of seventeen before he could read and then went to Glasgow to study carpentry with his brother-in-law. When the latter's business failed, in 1748 John was sent to London to stay with his older brother William, who ran an anatomy school. John was an immediate success, revealing an astonishing dexterity and skill in dissection. Before the end of his first year he was given the job of preparing the "subjects" (usually the bodies of executed criminals) for each dissection lesson. After eleven years with William, John spent three years as an army surgeon and wrote *A Treatise on Blood, Inflammation and Gunshot Wounds*. He then married Anne Home (who wrote librettos for the composer Haydn) and published the first scientific work on the treatment of teeth. In 1774 Hunter became a director of the Humane Society,[133] set **133** 5 *28*
up a highly successful practice and began to indulge in his various hobbies. These included hedgehogs, the structure of whales and codfish hearing. In 1759 the Hunter anatomy school had taken on a new pupil, named William Hewson, and in 1762 when John retired because of ill-health, Hewson, who had attended William's lectures and lodged with John, took over as assistant and then as William's partner.

In 1774 Hewson was to die from an infection, four years after marrying a woman named Mary Stephenson. In the 1750s and 1760s Mary had taken in lodgers, one of whom, on two separate occasions, was Benjamin Franklin.[134] He had been in England on the first occa- **134** 15 *38*
sion, from 1757, to argue the right of Pennsylvania to raise certain taxes and on the second (and more historic) visit to argue against the British right to tax the American colonies without granting them representation in Parliament. After failing in this second endeavor

Franklin returned to America and became a leading figure in the events leading up to the Declaration of Independence in 1776. In the years that followed William Hewson's death Franklin tried again and again to persuade Mary Hewson to come to America to be with him, on one occasion writing: "Your joining me . . . will surely make me happier, provided your change of country may be for the advantage of your dear little family. When you have made up your mind on the subject, let me know by a line, that I may prepare a house for you as near me, and otherwise as convenient for you as possible." In 1786 Mary finally went to Philadelphia and spent three and a half years looking after Franklin until he died.

Franklin gained a deservedly international reputation for his research into the nature of electricity. Work for which he is less well-known was related to the numerous occasions on which he crossed the Atlantic either to England before Independence or later as U.S. minister to France. In 1769 Franklin had heard rumors that the fast ships on which transatlantic mail was dispatched did well en route from America to Europe but were unaccountably slow on the return, taking up to two weeks longer to complete the journey. When relations between America and England became more strained in the period immediately before independence this postal delay became a critical matter. Franklin sought the advice of Captain Timothy Folger, a relative of his mother who had for years commanded whaling ships out of Nantucket. Folger told him of a mysterious "river of the sea" that whalers knew about, which ran up the East Coast of America and then off toward Europe. The whalers used this current to enhance their speed going east and on their return zig-zagged back and forth across it to avoid delay.

On several of his transatlantic trips after 1775 Franklin investigated the mysterious current. From early morning to late evening every day as he crossed the ocean he measured the temperature of the water in and around the current by lowering a corked bottle to below 210 feet, where the pressure forced in the cork and the bottle filled. The bottle was then quickly drawn up and the water temperature measured. In this way, Franklin was able to trace the boundaries of the current whose water turned out to be up to six degrees Fahrenheit warmer than the surrounding sea. Thanks to this temperature profile Franklin was able to map the current. His chart was the first detailed map of the Gulf Stream.

Franklin took his temperature profile of the Gulf Stream in Fahrenheit degrees because by then the Fahrenheit scale was in fairly general use. This had not long been the case. As late as mid-eighteenth century it was not uncommon for there to be as many as a dozen different temperature scales. But at a time when science and technology required ever more precision this disorganized state of affairs was not helpful. The problem was solved by Daniel Gabriel Fahrenheit, an instrument-maker born in Danzig and sent to learn business in Amsterdam. In 1707 at the age of twenty-one Fahrenheit left the city for ten years of travel around Europe, visiting other instrument-makers and scientists, first in Germany and then (in 1708) in Denmark. In Copenhagen he met the ex-mayor, a talented scientific amateur named Ole Roemer, and was able to watch him at work. In 1717 when Fahrenheit returned to Amsterdam to set up as an instrument-maker he took with him the notes he had made of Roemer's work.

Using a mercury thermometer, Roemer had taken the temperature of the armpit of a healthy male and marked the level of the mercury. He then marked the spot to which the mercury fell when the thermometer was immersed in a freezing water. Since the temperature of a mixture of salt and ice was considered to be the coldest possible, he called this zero. Fixing an upper limit (boiling water) at 60, freezing water came at 7.5 (one-eighth up the scale), and a healthy armpit came at 22.5 (three-eighths up the scale).

Fahrenheit decided in the interest of greater precision to expand Roemer's scale, multiplying each number by four. This set freezing at 30 and armpit temperature at 90. To eliminate the awkward fractions this would cause and to maintain divisibility by eight, he moved the freezing point to 32 and armpit temperature to 96. The later, minor change of armpit (blood) temperature to 98.6 set the modern thermometer scale, which we attribute to Fahrenheit because Ole Roemer's own notes were destroyed in a fire and little was known of his thermometer work for two hundred years.

However, Roemer was known at the time for other, more cosmic
matters. Earlier, in 1671, while he was still involved in astronomical
studies, a passing French astronomer had persuaded him to work as
135 107 *206* his assistant during a visit to the Danish astronomer Tycho Brahe's[135]
observatory at Uraniborg on the island of Hven (between Denmark
and Sweden). The purpose of the visit was to check the observatory's
exact coordinates as part of a major French program to update astronomical tables such as those Brahe himself had produced. Roemer
then went to Paris and spent several years developing an extraordinary idea that had occurred to him while carrying out observations
on Hven.

One of the exact coordinates in the sky Roemer had measured for the French was the position and time of eclipse of Io, one of Jupiter's moons. Precise moments of celestial time such as these were of great value to mariners, as they allowed exactness in calculating longitude position at sea, which is determined by the time a celestial event happens compared with when it would occur at home port. The difference between the two times at which the phenomenon was observed told a navigator how far east or west he was. It was during the observations of Io that Roemer began to wonder why the time at which the eclipses occurred varied with the distance between Jupiter and the Earth. He came to the momentous conclusion that these differences must be due to the fact that the speed of light was finite (and not instantaneous, as had been believed ever since the time of Aristotle). The image of the eclipse must therefore be taking longer to arrive at Earth the greater the distance from Jupiter. Calculations based on this assumption led Roemer to announce on November 21, 1676, that the speed of light was 140,000 miles per second.

During the previous two years the man who had persuaded Roe-
136 101 *188* mer to come to Paris, fellow-astronomer Jean Picard,[136] was concerned with more down-to-earth matters. Between 1674 and 1675 he
played an active role in assuring a constant water supply for the
137 89 *161* king's new château at Versailles.[137] In 1671 Louis XIV had started to
build a great royal palace with magnificent gardens on the site of his

father's relatively modest hunting lodge. The construction of Versailles and its gardens took thirty-six thousand workers twenty-six years.

Picard's water problem related to the fact that Louis wanted the latest fountains and water-powered amusements in grottoes all over the palace grounds. Water was also needed for the hundreds of thousands of plants, trees and shrubs in the gardens The hydrological difficulty was that Versailles turned out to be higher than the surrounding ground. This awkward fact was revealed by Picard, who used an adaptation of his astronomical telescope to survey the different levels with extreme accuracy. Thanks to this a complex network of canals and aqueducts was subsequently designed and brought water from reservoirs and springs in the surrounding countryside. From 1683 on, the Versailles gardens were being adequately supplied.

This was good news to André Le Nôtre, the gardens' designer, who must have been an exceptional person, since he was described by contemporaries as "honest, honorable and plain-spoken." It was even said that Le Nôtre achieved an almost personal relationship with his sovereign, the Sun King. Le Nôtre's Versailles gardens, built for an absolute monarch and unequalled in their size and complexity, were intended to reflect the concept of the king's supreme control. At a time when exploration was opening up the world and science was revealing the secrets of the cosmos, Versailles represented another aspect of man's newfound power. Untamed countryside no longer surrounded society with mysterious and uncontrollable chaos as it had done since medieval times. Elegant lines of trees and carefully delineated flowerbeds now shaped and constrained the elements. The king also controlled nature.

His chief minister wanted that control to extend further. In the late seventeenth century Jean-Baptiste Colbert[138] was busy trying to re- **138** 84 *159*
trieve the French economy from the near-bankrupt state in which Louis XIII had left it. As part of his plan (which included revamping the whole of French industry so that France would no longer need to import goods), Colbert made a complete reorganization and reform of the French navy. Colbert's dream was that France would become a

world power ranking with England. His ambitious ship-building program included the introduction of draconian afforestation laws. For centuries the woods had been decimated by charcoal-burners and fuel-collectors. The new regulations now preserved the trees for ship-builders alone (one side-effect would be to drive the iron industry to seek other sources of fuel and ultimately to develop ways of using coal that would power the Industrial Revolution).

It was thanks to the continuing scarcity of wood that the inspector general of the French Navy in 1732 was an already-recognized botanist. Henri-Louis Duhamel du Monceau had begun life as a chemist, but after a visit to England in 1729 to study ship-building concentrated much of his attention on wood and forest management. His first book, published in 1747, was about ships' rigging. At the family château of Denainvilliers, between Orleans and Chartres, Duhamel experimented with the latest English agricultural techniques and started one of Europe's first arboreta, gathering specimens from all over the European continent and from America. Duhamel's treatise on trees and shrubs was widely influential in the early importation of new plant species. In 1750 he translated the

Fig. 32: *Duhamel's legacy. An eighteenth-century French print from a book on forest management.*

English agricultural expert Jethro Tull's seminal work on *Horse-Hoeing Husbandry,* adding to it from his own experience. Earlier in the century Tull had seen French peasants hoeing their vineyards, and when he used the technique in England found he could produce a wheat crop from the same ground for thirteen consecutive years without the need for expensive manure. Ironically, this French innovation publicized by an Englishman now formed the basis of Duhamel's *Traité de la Culture des Terres* ("Treatise on Cultivation"), which was in turn translated into English in 1759 as *A Practical Treatise of Husbandry* by John Hill, assistant gardener at the new Royal Botanical Gardens in Kew, London. Hill was also the author of a catalogue of some thirty-four hundred species of plants being cultivated in Kew at the time.

In 1761 an English architect and author named William Chambers asked his publisher to send him a list of the latest books on gardening (likely including Hill's work and his translation of Duhamel), because Chambers had just been commissioned by the dowager princess Augusta, King George III's mother, to carry out architectural work at Kew. Chambers was already famous for his books, *Treatise on Civil Architecture* and *Designs for Chinese Buildings,* which had led to his earlier appointment as architecture tutor to the Prince of Wales, who was now the king. Chambers had learned about Chinese style between 1742 and 1749 when he was in China working for the Swedish East India Company. These voyages convinced him to quit commerce, and in 1749 he moved to France and Italy for six years to study architecture.

In response to Princess Augusta's commission Chambers designed more than twenty buildings for Kew Gardens, chief among which (and still standing) was the Pagoda, an octagonal ten-story structure 163 feet high, more of a rococo statement than a true Chinese reproduction. The building caused a sensation and set the fashion for architectural *chinoiserie.* Pagodas popped up in parks all over Europe in Potsdam, Munich, Tsarskoie Selo, Chanteloup and Oranienbaum. Thanks to his links with royalty, Chambers was appointed architect of the works together with Robert Adam, and in 1784 he became surveyor-general, titular head of the architectural profession in England. In 1774 he and Adam were commissioned to work on Somerset House, a grandiose new public building in central London. In 1782 the two men hired a Scottish stoneworker, Thomas Telford, who would later describe Chambers as "haughty and reserved."

Telford was a self-taught journeyman stonemason with ambitions to be an architect and planner, and he spent two years working at Somerset House, learning all he could. After leaving Chambers and Adam, Telford built a dockyard house, remodeled a castle, designed a prison, a church and a hospital and became Shropshire's county surveyor. His civil engineering career began when he was appointed to work on a network of new canals linking the rivers Dee, Mersey and Severn. The aqueduct he built for this project at Pontcysyllte in Wales is one of the engineering triumphs of history. The canal is carried one thousand feet across a wide, deep valley in a cast-iron trough eleven feet ten inches wide, containing the aqueduct and the tow-path and supported on nineteen slender stone pillars, each 127 feet high. Sir Walter Scott called it the greatest work of art he had ever seen. Even today it is impressive and beautiful.

In 1801 Telford was commissioned to build the great Caledonian Canal through the Highlands of Scotland. During the eighteen years this took to complete Telford also built 920 miles of new roads and over one thousand new bridges, transforming the Scottish economy by making possible stagecoach services, regular mail and newspaper delivery. This in turn boosted commercial activity, which increased land and property values. By 1820 Telford was the first president of the new Institute of Civil Engineers. When he died in 1834, popular and admired, he was buried in Westminster Abbey.

His only failure had been the design he entered for the competition to build a new London bridge. Telford's idea was for a single, six-hundred-foot, cast-iron span rising to sixty-five feet above the river, with a roadway forty-five feet wide, and weighing six thousand tons. In 1816 when the government consulted a group of eminent scientists and engineers on the subject their opinion was that the bridge was brilliantly conceived. Unfortunately, in order to avoid too great a rise at the crown of the arch (necessary to facilitate the passage of shipping), Telford planned high-level approach ramps supported by brick colonnades running along either bank of the river. The cost of the land on which these ramps would be built was too great and Telford's design was rejected.

A member of the ad hoc bridge review group (which also in-
cluded James Watt[139] and ironmaster James Wilkinson) was Thomas **139** 16 *39*
Young,[140] one of the most innovative and versatile scientists of the **139** 37 *68*
day. Young was a prodigy. By the age of two he was literate and at four **140** 91 *163*
he had read the Bible twice. By the age of nineteen he was proficient in twelve languages, ancient and modern. He had also mastered calculus and read Newton's *Principia Mathematica* and *Optics,* and Lavoisier's *Elements of Chemistry.* At this point he began studies to qualify as a physician and then attended Cambridge, where he was known as "Phenomenon" Young. In 1801 he was appointed professor at the Royal Institution in London, where his work was to prepare popular lectures on science and the mechanical arts. He also made major advances in the study of color vision and perception, and in deciphering Egyptian hieroglyphics.

In 1799 he had also begun the study of light, and in 1807 he published a paper describing a series of experiments in which he had shone candlelight through a lens, then a pinhole, and finally two narrow slits. Beyond the slits, where the light fell on a sheet of paper, Young saw a series of light-and-dark patterns that he concluded must have been produced when the light coming though the slits recombined. Since the effect of this recombination was to produce what appeared to be interference patterns similar to those produced by interacting ripples of water, Young announced that contrary to all contemporary opinion he believed light traveled in waves through some kind of "luminiferous ether."

This ether was of course invisible and intangible (and ubiquitous, since light also traveled in a vacuum). The search for this mysterious "ether" would bedevil the work of scientists for most of the nineteenth century. In 1888 a German physicist named Heinrich Hertz conducted a series of experiments to see if electromagnetic radiation traveled through the ether as did light waves. Confirmation came with his discovery of radio waves. Hertz had been directed to this research by his professor at the University of Berlin, Hermann von

Helmholtz, a leading figure in European science. Helmholtz had studied physiology under Johannes Müller, who wrote the *Handbook of Physiology,* a milestone in the history of European medicine. Müller (and then Helmholtz) rejected earlier, quasiphilosophical approaches to physiology based on speculative Romantic thought, relying instead on the empirical evidence of observation and experimentation. One of Müller's greatest contributions to neurophysiology was to encourage the view of the nervous system as a unit.

Strangely enough, it was in matters such as nerve function that Müller clung to a view that seemed to contradict his belief in empirical evidence. Müller was a Vitalist, holding that it would be impossible to reduce the life processes to the mechanical laws of chemistry and physics. Vitalists believed that the organism as a whole was greater than the sum of its parts and that some kind of "force" coordinated the physiological action of organs, nerves and tissues to produce the harmonious behavior that characterized an organism. Vitalists held that this force was not susceptible to experimental quantification. Müller used the nerve impulse as an example.

His pupil Helmholtz disagreed violently with the whole Vitalist approach and set out to disprove it. In 1852 he published evidence from experiments he had conducted on a frog's sciatic nerve. Using a myograph, Helmholtz recorded the effect of sending an electric current into the nerve. The muscle contraction that followed nerve excitation was recorded with a lever that moved with the muscle spasm and traced a curve on a smoked-glass surface moving at a uniform speed. The vertical component of the curve was proportional to the contraction of the muscle and the horizontal component was proportional to time. Helmholtz found not only that the nerve impulse occurred in a finite period but also that it moved at the relatively slow speed of about ninety feet per second.

Vitalists ignored the findings. By 1900 the leader of the German Vitalist movement was a physicist, chemist and philosopher named Ludwig Klages, who was a disciple of Nietzsche and a devotee of antirationalism. For Klages the intellect was a superimposed power, constraining the naturally intuitive and "prophetic" mind. In his work over and over he exhorted psychologists to turn away from rationalism toward "divination." In Monaco in 1905 Klages founded a center for the study of "Characterology," an alternative form of psychology that assessed personality through what today would be referred to as "body language." Klages and his followers claimed that intuitive analysis of the conflicting elements of a person's character as shown by movement and expression would reveal what lay behind what he called "the mask of courtesy."

In an extension of this approach to the study of character (which together with characterology he offered as a tool for personnel selection), Klages also published a book on graphology, the study of handwriting. The book was a runaway success and went into fifteen editions. Klages maintained that handwriting offered added insights into character, since it was affected by the different driving forces in the personality. Thus, for instance, large handwriting indicated either enthusiasm or a lack of realism, whereas small handwriting indicated either lack of enthusiasm or realism. Sloping writing reflected either congeniality or rashness. Vertical writing indicated either rationality or aloofness. The choice of either the positive or negative characteristic in each case depended on the rhythm, depth and richness of the script. This final feature could, however, not be measured but only intuitively perceived. Given the irrational element in Klages's thought, it is not surprising that his characterology and graphology were most extensively used by the Nazis to select their SS officers.

Klages's graphology drew attention to the extreme individuality of each person's handwriting. Thirty years after the Nazis had adopted his theories the extreme individuality of handwriting was presenting problems to the United States Post Office Department, at the time facing an unmanageably massive growth in business mail. By the early 1960s it accounted for 80 percent of all correspondence, and the amount of mail was growing rapidly. The single greatest contribution to this growth had come from the introduction of the computer. In permitting the centralization of accounts it generated a vastly increased amount of billing, bank deposits and receipts, credit card transactions, Social Security payments and other business items traveling through the postal system. There was an urgent need to streamline the process of sorting and delivery. On July 1, 1963, the

Post Office Department introduced the five-digit ZIP ("Zone Improvement Plan") Code. The first number in the code indicated a broad geographical area (for instance, zero referred to the Northeast and nine to the West); the following two digits referred to areas of population concentration or areas using a common transportation system; the final two digits referred to small post offices or postal zones in larger, zoned cities. In the 1980s the U.S. Postal Service added four more digits, permitting sorting and delivery to be automated to the level of an individual building.

Automatic processing of mail began in 1965 with the installation at the Detroit Post Office of a high-speed Optical Character Reader. This first-generation machine read the line of typed or block-capital letters carrying the name of the city and the state ZIP Code, then sorted mail into one of 277 pockets. However, each address had to be checked by an agent before delivery. In the 1980s more sophisticated machines were developed capable of reading an individual ZIP Code and spraying the relevant machine-readable barcode on the mail, which could then be sorted by computer. At the end of the century, still constrained by what Klages had called the "extreme individuality" of handwriting, even the most advanced, "multiline" OCRs were still only capable of reading block-letter handwriting or typed letters.

Some of the earliest research in optical character recognition
started in 1952 with research into systems to assist reading by the
blind carried out by the Cognitive Information Processing Group at
141 2 23 MIT. This early research shared a common origin in cybernetic[141]
feedback theory with the guidance technology described at the beginning of this book's journey on the web of knowledge: the elec-
142 1 22 tronic personal agents[142] that will keep us in touch with the world in
the twenty-first century.

Bibliography

CHAPTER 1

Bradley, Ian. *William Morris and His World.* London: Thames & Hudson, 1978.

Carson, Gerald. *Cornflake Crusade.* London: Victor Gollanz Ltd., 1959.

Corbin, Diana Fontaine Maury. *A Life of Matthew Fontaine Maury.* London, 1888.

Guilfoyle, Christine, and Warner, Ellie. *Intelligent Agents.* London, 1994.

Hibbert, Christopher. *Nelson.* London: Viking, 1994.

Holmes, Frederic Lawrence. *Claude Bernard and Animal Chemistry.* Cambridge, Mass: Harvard University Press, 1974.

Jameson, Eric. *The Natural History of Quackery.* London: Michael Joseph, 1961.

Masani, P. R. *Norbert Wiener.* Basel, Boston, Berlin: Birkhäuser Verlag, 1990.

Spencer, Colin. *The Heretic's Feast.* London: Fourth Estate, 1993.

Sultana, Donald. *Samuel Taylor Coleridge in Malta and Italy.* Oxford: Basil Blackwell, 1969.

Taylor, Anne. *Annie Besant.* Oxford: Oxford University Press, 1992.

CHAPTER 2

Batty, Peter. *The House of Krupp.* London: Secker and Warburg, 1966.

Dibner, Bern. *The Atlantic Cable.* New York: Blaisdell, 1964.

Hyman, Anthony. *Charles Babbage.* Princeton: Princeton University Press, 1992.

Leupp, F. E. *George Westinghouse.* Norwood, Mass.: Norwood Press, 1919.

Mackenzie, Thomas B. *Life of James Beaumont Neilson, F.R.S.* Glasgow: West of Scotland Iron and Steel Institute, 1928.

McLaren, David J. *David Dale of New Lanark.* Milngavie: Heatherbank Press, 1983.

Mitchell, F. *Tank Warfare*. Stevenage, Herts.: Spa Books, 1987.
Moon, John F. *Rudolf Diesel and the Diesel Engine*. London: Priory Press Ltd., 1974.
Rosenberg, Nathan. *The Britannia Bridge*. Cambridge, Mass.: MIT Press, 1978.
Seeligman, T., Torrilhon, G., and Falconnet, H. *Indiarubber and Gutta Percha*. London: Scott, Greenwood & Sons, 1910.
Stigler, Stephen M. *The History of Statistics*. Cambridge, Mass.: Harvard University Press, 1986.
Vaughan, Adrian. *Isambard Kingdom Brunel*. London: John Murray, 1991.

CHAPTER 3

Baines, Edward. *History of the Cotton Manufacture*. New York: Augustus M. Kelley, 1966.
Blackmore, John. *Ernst Mach*. London: University of California, 1973.
Carmichael, Leonard, and Long, J. C. *James Smithson and the Smithsonian Story*. New York: G. P. Putnam's Sons, 1965.
Crocker, Glenys. *The Gunpowder Industry*. Haverfordwest: Shire Publications Ltd., 1986.
Gernsheim, Helmut, and Gernsheim, Alison. *L. J. M. Daguerre*. London: Secker & Warburg, 1956.
Hackmann, Willem. *Seek and Strike*. London: H.M.S.O., 1984.
Harris, Neil. *Humbug, the Art of P.T. Barnum*. Chicago: University of Chicago Press, 1981.
Lewes, Vivian B. *Acetylene*. London: Macmillan, 1900.
Pannekoek, A. *A History of Astronomy*. London: George Allen & Unwin Ltd., 1961.
Phillips-Matz, Mary Jane. *Verdi*. Oxford: Oxford University Press, 1993.
Quinn, Susan. *Marie Curie*. London: Heinemann, 1995.
Sawyer, L. A. *Liberty Ships*. New York: Lloyd's of London, 1985.
Style, Jane M. *Auguste Comte*. London: Kegan Paul, Trench Truebner & Co., Ltd., 1928.
White, Michael, and Gribbin, John. *Einstein*. London: Simon & Schuster, 1994.

CHAPTER 4

Blackstone, Sarah J. *Buckskins, Bullets, and Business*. Westport, Conn.: Greenwood Press, 1986.

Bortoloan, Liana. *The Life and Times of Titian.* London: Hamlyn Publishing Group, 1968.

Bradford, Ernle. *The Great Siege.* London: Hodder & Stoughton, 1961.

Cluny, Hilaire. *Louis Pasteur.* London: Souvenir Press, 1965.

Hatch, Alden. *American Express.* Garden City, N.Y.: Doubleday & Co., Inc., 1950.

Main, Gloria L. *Tobacco Colony.* Princeton: Princeton University Press, 1982.

Neillands, Robin. *The Hundred Years War.* London & New York: Routledge, 1990.

O'Malley, Charles D. *Andreas Vesalius of Brussels.* Berkeley and Los Angeles: California Press, 1964.

Ordish, George. *The Great Wine Blight.* London: J. M. Dent & Sons Ltd., 1972.

Sharov, Alexander S., and Novikov, Igor D. *Edwin Hubble.* Cambridge: Cambridge University Press, 1993.

Shaw, Stanford J. *The Jews of the Ottoman Empire and the Turkish Republic.* London: Macmillan, 1991.

Shennan, Francis. *Flesh and Bones, The Passions and Legacies of John Napier.* Edinburgh: Napier Polytechnic, 1989.

Smith, Melvyn. *Space Shuttle.* Sparkford, Somerset: Haynes, 1989.

Warner, Marina. *Joan of Arc.* London: Weidenfeld & Nicholson, 1981.

CHAPTER 5

Bull, Angela. *The Machine Breakers.* London: Collins, 1980.

Clayton, Michael. *The Jeep.* London: David & Charles, 1982.

Clifford, M. N., and Wilson, K. C. (eds.). *Coffee.* London: Croom Helm, 1985.

Goldsmith, Margaret. *Christina of Sweden.* London: Arthur Baker Ltd., 1933.

Gordon, Alistair. *John Galt.* Edinburgh: Oliver & Boyd, 1972.

Grendel, Frédéric. *Beaumarchais.* Trans. Roger Greaves. London: Macdonald and Jane's, 1977.

Hughes, J. Trevor. *Thomas Willis.* London: Royal Society of Medicine Services Ltd., 1991.

Hutchison, Harold F. *Sir Christopher Wren.* London: Victor Gollancz Ltd., 1976.

Lande, Dr. Lawrence. *Introduction to John Law.* Edinburgh: University of Edinburgh Centre for Canadian Studies, 1989.

Longford, Elizabeth. *Byron.* London: Hutchinson, 1976.

Oppenheimer, Jane M. *Essays in the History of Embryology and Biology.* Cambridge, Mass.: M.I.T. Press, 1967.

Parker, Geoffrey. *The Military Revolution.* Cambridge: Cambridge University Press, 1988.

Pattison, Mark. *Isaac Casaubon.* Oxford: Clarendon Press, 1982.

Robinson, George W. *Autobiography of Joseph Scaliger.* Cambridge, Mass.: Harvard University Press, 1927.

Winegarten, Renée. *Mme de Staël.* Leamington Spa: Berg Publishers Ltd., 1985.

CHAPTER 6

Aitken, Hugh G. *Syntony and Spark—The Origins of Radio.* New York and London: John Wiley & Sons, Inc., 1976.

Barchilon, Jacques, and Flinders, Peter. *Charles Perrault.* Boston: Twayne Publishers, 1981.

Campbell, Malcolm. *Pietro da Cortona at the Pitti Palace.* Princeton: Princeton University Press, 1977.

Houghton, Raymond. *The World of George Berkeley.* Dublin: Eason & Son Ltd., 1985.

John, William D. *Pontypool and U.K. Japanned Wares.* Newport, Monmouthshire: The Ceramic Book Co., 1953.

Koepke, Wulf. *Johann Gottfried Herder.* Boston: Twayne Publishers, 1987.

Lavine, Sigmund A. *Allan Pinkerton.* London: Mayflower Paperback, 1970.

May, Stacy. *United States Business Performance Abroad: The Case Study of the United Fruit Company in Latin America.* New York: National Planning Association, 1958.

Nordon, Pierre. *Conan Doyle.* London: John Murray, 1966.

Rolt, L. T. C. *The Aeronauts, A History of Ballooning.* Gloucestershire: Alan Sutton, 1985.

Rowse, A. L. *Jonathan Swift, Major Prophet.* London: Thames & Hudson, 1975.

Smith, Charles H. *A. F. Wallace on Spiritualism, Man and Evolution.* 1992.

Stafford, Fiona. *The Sublime Savage. A Study of James Macpherson and the Poems of Ossian.* Edinburgh: Edinburgh University Press, 1988.

Stanton, Phoebe. *Pugin.* London: Thames & Hudson, 1971.

Vining, Elizabeth Gray. *Flora MacDonald.* London: Geoffrey Bles, 1967.

CHAPTER 7

Allan, D. G. C., and Schofield, R. E. *Stephen Hales*. London: Scholar Press, 1980.

Burke, Peter. *Montaigne*. Oxford: Oxford University Press, 1994.

Chancellor, John. *Audubon: A Biography*. London: Weidenfeld and Nicolson, 1978.

Dickinson, Robert E. *Makers of Modern Geography*. London: Routledge and Kegan Paul, 1969.

Fisher, Richard B. *Edward Jenner*. London: André Deutsch, 1991.

Hartcup, Adeline. *Angelica*. Harmondsworth, Middlesex: William Heinemann Ltd., 1954.

Kidler, Peter. *Richthofen*. London: Arms + Armour Press, 1994.

Koestler, Arthur. *The Watershed: A Biography of Johannes Kepler*. Lanham, Md: University Press of America, 1960.

Mackworth-Praed, Ben. *Aviation: The Pioneer Years*. London: Studio Editions Ltd., 1990.

Marriott, Ernest G. *Izaak Walton*. Nottingham: Nottingham Flyfishers' Club, 1986.

Morley, Geoffrey. *The Smuggling War*. Stroud: Alan Sutton Publishing Ltd., 1994.

Roddis, Louis H. *James Lind*. London: William Heinemann Ltd., 1951.

Stone, George W., and Kahrl, George M. *David Garrick*. Southern Illinois University Press, 1979.

CHAPTER 8

Blumenberg, Hans. *The Genesis of the Copernican World*. Trans. Robert M. Wallace. Cambridge, Mass.: M.I.T. Press, 1987.

Boase, Roger. *The Origin and Meaning of Courtly Love*. Manchester: Manchester University Press, 1977.

Craven, William G. *Giovanni Pico della Mirandola*. 1981.

Crone, Gerald. *Maps and Their Makers*. London: Hutchinsons University Library, 1953.

Dan, Joseph. *The Early Kabbalah*. Mahwah, N.J.: Paulist Press, 1986.

Dodge, Ernest Stanley. *The Polar Rosses*. London: Faber & Faber, 1973.

Gade, John Allyne. *The Life and Times of Tycho Brahe*. New York: Princeton University Press, 1947.

Holmes, T. W. *The Semaphore*. Ilfracombe, Devon: Arthur H. Stockwell Ltd., 1983.

Kardross, John. *The Origins and Early History of Opera.* Sydney: University of Sydney Press, 1957.

Manschreck, Clyde Leonard. *Melanchthon.* New York: Oxford University Press, 1970.

Ore, Oystein. *Cardano.* Princeton: Princeton University Press, 1953.

Schwarzbach, Martin. *Alfred Wegener.* Madison, Wisc.: Science Technology Inc., 1986.

Spitz, Lewis W. *The Religious Renaissance of the German Humanists.* 1963.

Turrill, William B. *Joseph Dalton Hooker.* London: Thomas Nelson & Sons Ltd., 1963.

Wormald, Jenny. *Mary Queen of Scots.* London: George Philip, 1988.

CHAPTER 9

Beattie, Lester M. *John Arbuthnot.* Cambridge, Mass.: Harvard University Press, 1935.

Besterman, Theodore. *Voltaire.* Oxford: Basil Blackwell, 1976.

Brown, Pamela. *Henri Dunant.* Dublin: Wolfhound Press, 1991.

Everdingen, Ewoud van. *C.H.D. Buys Ballot.* Antwerp, 1953.

Halsband, Robert. *The Life of Lady Mary Wortley Montagu.* New York: Oxford University Press, 1960.

Killingray, David. *The Atom Bomb.* London: Harrap, 1983.

Lawson, Joan. *A History of Ballet and Its Makers.* London: Sir Isaac Pitman & Sons Ltd., 1964.

Lumsden, Malvern. *Incendiary Weapons.* Cambridge, Mass.: M.I.T. Press, 1975.

MacGregor, Arthur. *Sir Hans Sloane.* London: British Museum Press, 1994.

Miller, Edward. *Prince of Librarians.* London: The British Library.

Millington-Drake, Egen. *The Drama of Graf Spee and the Battle of the Plate.* London: Peter Davies, 1965.

Smith, Maxwell A. *Prosper Mérimée.* New York: Twayne Publishers, Inc., 1972.

Stephens, W. P. *Zwingli.* Oxford: Clarendon Press, 1994.

Stuyvenberg, J. H. van. (ed.). *Margarine: An Economic, Social and Scientific History.* Liverpool: Liverpool University Press, 1969.

Uerberhorst, Horst. *Friedrich Ludwig Jahn and His Time.* Munich: Heinz Moos Verlag, 1982.

CHAPTER 10

Bourde, André J. *The Influence of England on the French Agronomes, 1750–1789*. Cambridge: Cambridge University Press, 1953.

Cahan, David (ed.). *Hermann von Helmholtz and the Foundations of Nineteenth-Century Science*. Berkeley, Calif.: University of California Press, 1993.

Checkland, Sydney. *The Elgins*. Aberdeen: Aberdeen University Press, 1988.

Clark, Ronald W. *Benjamin Franklin*. London: Weidenfeld and Nicolson, 1983.

Cohen, Ernst Julius. Kammerlingh Onnes Memorial Lecture, in *Journal of the Chemical Society*, 1920. Vol. 1, pp. 1193–1209.

Cohen, I. B. "Roemer and the first determination of the velocity of light," in *Isis*, 31 (1940), pp. 327–79.

Dorsey, N. Ernest. "Fahrenheit and Roemer," in *Journal of the Washington Academy of Sciences*, 36, No. 11 (1946), 361–72.

Harris, John. *Sir William Chambers*. London: A. Zwemmer Ltd., 1970.

Harriss, Joseph. *The Eiffel Tower.* London: Paul Elek, 1976.

Hazlehurst, F. Hamilton. *Gardens of Illusion: The Genius of André Le Nostre*. Nashville, Tenn.: Vanderbilt University Press, 1980.

Levey, Michael. *Sir Thomas Lawrence*. London: National Portrait Gallery, 1979.

Qvist, George. *John Hunter.* London: William Heinemann Medical Books Ltd., 1981.

Rolt, L. T. C. *Thomas Telford*. Harmondsworth, Middlesex: Penguin Books Ltd., 1979.

Taylor, Anne. *Laurence Oliphant*. Oxford: Oxford University Press, 1982.

Vogel, Dan. *Emma Lazarus*. Boston: Twayne Publishers, 1980.

Index

Index

About the Author

James Burke, a noted authority on the history of technology and science, is the bestselling author of *Connections*, *The Day the Universe Changed* and *The Pinball Effect* and, with Robert Ornstein, of *The Axemaker's Gift*. He also hosts the highly rated documentary television series *Connections*. He lives in London, England.

Illustration Credits

AKG London: Figs. 5, 11, 20, 23, 27, 28.

Corbis U.K.: Figs. 8, 13, 14, 16, 29, 30.

Mary Evans Picture Library, London: Figs. 1, 3, 4, 6, 9, 10, 12, 15, 17, 19, 21, 22, 26.

National Portrait Gallery, London: Figs. 2, 18, 24, 25, 31.

Royal Horticultural Society, London: Fig. 32.

Royal Photographic Society, U.K.: Fig. 7.